Towards User-Centric Intelligent Network Selection in 5G Heterogeneous Wireless Networks

Zhiyong Du · Bin Jiang · Qihui Wu ·
Yuhua Xu · Kun Xu

Towards User-Centric Intelligent Network Selection in 5G Heterogeneous Wireless Networks

A Reinforcement Learning Perspective

 Springer

Zhiyong Du
National University of Defense Technology
Changsha, Hunan, China

Qihui Wu
Nanjing University of Aeronautics
and Astronautics
Nanjing, Jiangsu, China

Kun Xu
National University of Defense Technology
Changsha, Hunan, China

Bin Jiang
National University of Defense Technology
Changsha, Hunan, China

Yuhua Xu
Army Engineering University of PLA
Nanjing, China

ISBN 978-981-15-1122-6   ISBN 978-981-15-1120-2   (eBook)
https://doi.org/10.1007/978-981-15-1120-2

This Springer imprint is published by the registered company Springer Nature Singapore Pte Ltd.
The registered company address is: 152 Beach Road, #21-01/04 Gateway East, Singapore 189721, Singapore

Preface

With the fast development of wireless communications, users nowadays are no longer satisfied with just being connected but pursue excellent service experience. Accordingly, how to promote users' QoE (quality of experience) beyond traditional QoS (quality of service) has received great attention from academia and industry in recent years. To this end, this book provides a systematic study on user-centric optimization idea from the perspective of network selection in 5G heterogeneous networks, where (i) the features of user demand are analyzed and characterized for optimizing QoE, which, in turn, drives us to rethink the optimization process in resource management, and (ii) end users can play a more active role in improving QoE, e.g., user-initiated active and fine-grained network handoff or users pay additional fee to pursue better QoE. On the other hand, from a practical perspective, due to complex and dynamic user demand and wireless network environment, network selection is confronted with incomplete and dynamic information, resulting in complex decision-making issue. To tackle this challenge, reinforcement learning based approaches are introduced to realize intelligent online optimization. Different reinforcement learning algorithms are proposed for specific network selection scenarios and problems.

The core idea of this book is "user-centric optimization + reinforcement learning algorithm". Actually, we established this idea and followed it in recent years. We believe that it could also contribute to the research of further wireless communications. The main content of this book is the related research results of the past 5 years.

This book can be used as a reference book for researchers and designers in resource management of 5G and beyond networks. First, this book tries to reveal the fact that user-centric resource management should not only be a change on the optimization objective, e.g., from QoS to QoE, but also the additional constraints of user demand on the whole optimization process. This new perspective opens new design principles for serving personalized user demand on future wireless systems. Second, this book inspires us that, compared with the objective QoS, the subjective features of QoE have the potential to enable more efficient resource utilization, where further in-depth research is needed to exploit user demand features. Finally, this book provides application examples of machine learning or artificial intelligent for resource management, particularly, how to promote QoE with user in the learning loop. The limitations and opportunities on using machine learning techniques for enabling artificial intelligence empowered wireless communications are preliminarily explored and discussed.

Besides the hard work and contributions of the coauthors, I am particularly grateful for my family's support. My parent always supports my work and encourages me. In particular, my wife Lijie helped me a lot. Although her working task is heavy, she managed many affairs for our family. Without her contribution, I cannot imagine how can I continue my work. Finally, I would like to say sorry to my wife and our cute daughter, Rongrong. I spent most of my time on work but too little time with them.

This book was supported by the Natural Science Foundation of China under Grant 61601490 and the Natural Science Foundation for Distinguished Young Scholars of Jiangsu Province under Grant BK20160034.

Wuhan, China   Zhiyong Du
September 2019

Contents

Acronyms

AP	Access point
CNP	Coupled network pair
CT-MAB	Continuous-time multi-armed bandit
DCF	Distributed coordination function
D-ONES	Decoupled online network selection
DRL	Deep reinforcement learning
E-LIA	Enhanced local improvement algorithm
FDMA	Frequency-division multiple access
HWN	Heterogeneous wireless networks
ITU	International Telecommunication Union
JFI	Jain's fairness index
LIA	Local improvement algorithm
L-SON	Local self-organizing
LTE	Long-term evolution
MAB	Multi-armed bandit
MADM	Multiple attribute decision-making
MARL	Multi-agent RL reinforcement learning
MDP	Markov decision process
MOS	Mean opinion score
MQE	Mixed strategy QoE equilibrium
NAP	Network access point
NE	Nash equilibrium
NSI	Network state information
ONES	Online network selection
PNE	Pure strategy Nash equilibrium
PPQE	Perfect pure strategy QoE equilibrium
PQE	Pure strategy QoE equilibrium
PSNR	Peak signal-to-noise ratio
QoE	Quality of experience
QoS	Quality of service

RHC	Rate of handoff cost
RL	Reinforcement learning
RR	Round-robin
RRM	Radio resource management
RSS	Received signal strength
SCTP	Stream control transmission protocol
SINR	Signal-to-interference-plus-noise ratio
SLA	Stochastic learning automata
TDMA	Time-division multiple access
TE	Trial and error
UCB	Upper confidence bound
UE	User equipment
UHD	Ultra-high-definition
VLC	Visible light communication
VM-ONES	Virtual multiplexing online network selection
WLAN	Wireless local network

Chapter 1
Introduction

1.1 Network Selection in Heterogeneous Wireless Network

In recent years, with the fast development of mobile Internet, the global mobile traffic demand has shown an explosive growth trend. Due to limited system capacity, traditional single cellular network is difficult to guarantee the quality of transmission services for users. As one approach to coping with this challenge, the concept of heterogeneous wireless networks [1] (HWN) is proposed. The core idea is to deploy multiple types of wireless networks (different types of radio access technologies such as cellular network and wireless local network (WLAN); and different coverage ranges of small cells such as microcell, picocell, femtocell) simultaneously, to form overlapping coverage, thereby effectively increasing the capacity of the system and enhancing network coverage. Currently, with the high data transmission rate, flexible and low-cost deployment, WLAN is widely used as a supplement of cellular networks. LTE-A (long-term evolution advanced) standards have introduced small cells and one of the key technologies of 5G communication systems is to increase system capacity through layered, ultra-dense small cell deployment. Therefore, the coexistence of HWN will remain for a long time.

Considering the relative closeness and independence among heterogeneous networks, the differences in working frequency bands, physical layer technologies, MAC layer multiple access technologies, and upper layer protocols, effectively exploiting the potential of HWN still faces many challenges, one of which is network selection. Network selection focuses on selecting the most suitable access network for multimode terminals to achieve the network diversity gain of heterogeneous wireless networks. Its essential goal is to adapt the distribution of wireless network resources to the spatiotemporal distribution of actual services, to improve the quality of services and achieve efficient use of wireless resources. It is worth noting that the "network" in "network selection" of this book refers to a general network access point (NAP), which can be any available base stations of cellular networks and access points (APs) of WLAN.

© Springer Nature Singapore Pte Ltd. 2020

Z. Du et al., *Towards User-Centric Intelligent Network Selection in 5G Heterogeneous Wireless Networks*,
https://doi.org/10.1007/978-981-15-1120-2_1

Most current network selection/handoff standards are mainly based on signal strength. For example, mobile devices in WLAN associate AP according to received signal strength and the network handoff of cellular networks is initiated once the measured reference signal received power (RSRP)/reference signal received quality (RSRQ) gap between the serving cell and candidate cell exceeds some threshold. It is true that signal strength relates to transmission performance, but possible noise and interference are also measured and the effect of wire network part is not accounted for. Alternatively, performance-oriented network selection and handoff are required. However, although industry and standardization organizations have made a lot of efforts to network convergence, such as 3GPP's multiple technical specifications for supporting 3GPP network and WLAN interconnection network architecture, and the IEEE 802.21 media independent network switching service for 3GPP, 3GPP2, IEEE 802 wired and wireless series standards, the support of cross-access technology and cross-operator network convergence is still not widely used [2]. This means that there is a lack of support for comprehensive network performance information and network selection guidance. In other words, network selection is not an easy task, but must consider practical consideration and constraints.

1.2 Challenges

The commercialization of smart terminals with "multimode" radio interface enables fine-grained network selection. Meanwhile, users' pursuits of higher transmission performance can be the intrinsic driver for network selection. Therefore, network selection is still important in the future. However, as mobile communication keeps evolving toward the 5G era, network selection faces several challenges as follows.

- **Heterogeneous and ultra-dense networks**. As we have mentioned, the coexistence of different wireless networks will last for a long time. Especially, besides traditional 3G/4G and WLAN, 5G is expected to introduce new millimeter-wave communication access technology. On the other hand, the ultra-dense deployment of small cells associated with the cloud radio access networks would aggravate the complexity of networks [3].
- **Large number of terminals**. Cisco mobile visual networking index predicates that the number of mobile phones will continue to grow fast and reaches 5.5 billion in 2021, where more than 50% are smartphones [4]. In addition, according to IDC, the number of global mobile devices including Internet of Things will be approximately 25 billion in 2020.
- **Various traffic types**. Apart from traditional multimedia services, the rise of mobile Internet has spawned a variety of applications such as mobile social applications, online games, and so on. Many new traffic and applications will continue to emerge such as virtual reality and augmented reality. These traffic types show different requirements on throughput, delay, and other performance metrics and there is a tendency that their requirements are becoming increasingly high.

- **Personalized user demands**. The network performance in the 3G/4G era is mainly based on the objective quality of service (QoS), such as throughput, while the subjective quality of experience (QoE) [5] has received great attention in recent years [6]. Notedly, as the traffic types become diverse, the user demand becomes more complex. Hence, network selection must meet context-aware even personalized user demands.

These new features in networks and terminals indicate that the network selection in 5G faces a more complex and dynamic external environment. The features in the traffic and user demands mean that more personalized and uncertain user demands should be considered in network selection. In summary, for the considered scenarios and features, achieving fine-grained and dynamic matching between differentiated network characteristics and diverse transmission requirements faces great challenges. Therefore, it is particularly important to optimize the network selection decision-making, which is the focus of this book.

1.3 Brief Review of Related Work

We give a brief review of previous works on network selection mainly from two aspects: evaluation criterion of network performance and network selection decision-making approach. The former focuses on how to evaluate and compare network performance for network selection. The latter provides a network selection decision with theory and tool basis.

There is a clear evolution history on the research of the evaluation criterion of network performance. Generally, from the physical layer to higher layer information, we can classify criteria into four types: received signal strength (RSS), QoS parameter, utility function, and QoE. RSS is highly related to wireless signal quality and can reflect distance information between transmitter and receiver, which is important for searching a network before access and handoff for mobile scenarios. Actually, RSS is the default AP selection criterion in some IEEE 802.11 specification. The main concern about RSS is that it cannot well reflect the transmission performance of wireless link. Another RSS-related criterion is signal-to-interference-plus-noise ratio (SINR), which is commonly used in load balancing [7] for ultra-dense small cells. To account for transmission capability, key QoS parameters [8] such as bandwidth, throughput [9], delay, are widely used in network selection. To characterize the relationship between QoS and traffic or application requirement, utility theory [10] has been introduced for evaluating the overall network performance. However, as QoE has received increased attention, how to bridge the gap between utility and QoE becomes a hot topic. To account for the subjective preference, standardization bodies has conducted serial works on audio, video, and audiovisual QoE models [11]. QoE has been used in network selection in [12] and more in-depth research is needed.

There are mainly the following five types of decision-making approaches in network selection. The basic approach is multiple attribute decision-making (MADM), which is suitable for the case with multiple parameters involved in network performance evaluation. To simulate human's decision-making, fuzzy logic is used to tackle the uncertainty and fuzziness in network selection parameters in [13]. Another powerful approach is neural networks. They is able to build a complex relationship between network performance parameters and direct network selection decision based on experience samples [12, 14]. On the other hand, when the dynamics in network performance and environment is considered, Markov decision process (MDP) could be adopted [15]. The advantage of MDP is that it provides a analytic framework to achieve long-term optimization in network selection and tradeoff handoff for dynamic scenarios. Finally, game theory is a powerful tool in analyzing multiple users' network selection behavior. Various game models have been introduced in user-controlled [16], network-controlled [17], and hybrid network selection models [18]. A survey of game-based network selection research can be found in [19, 20].

1.4 Main Idea

This book presents recent advances on network selection with a special perspective: User-centric optimization and reinforcement learning based solutions. The motivations for these two aspects are explained in the following.

1.4.1 User-Centric Optimization

The ultimate goal of wireless networks is serving users with different demands. Nowadays, users are no longer satisfied with just being connected, but pursue excellent service experience. Accordingly, how to promote users' QoE beyond traditional QoS has received great attention from academia and industry in recent years [21]. Instead of devoting to specific QoE metrics and optimization methods, this book provides a systematic study on user-centric optimization idea, from the perspective of access network selection. The user-centric optimization can be understood from the following two aspects.

- **The features of user demand are analyzed and characterized for optimizing QoE.** Due to the subjective features and complex model of user demand, there is an inherent gap between QoS (such as throughput) and QoE as illustrated in Fig. 1.1, Thus, traditional "QoS-oriented" approaches may result in two undesired situations: over-supply (QoE improvement is achieved at the expense of excessive resource provision) and under-supply (user demand cannot be fully met although targeted QoS is achieved). While QoE has been used as objectives of many current resource management issues, this book goes beyond by considering QoE con-

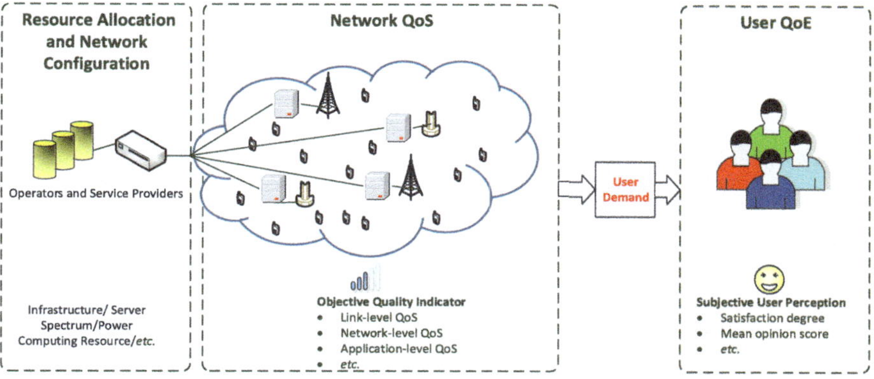

Fig. 1.1 QoE and QoS in network resource management [23]

straints on problem modeling and algorithm design, and exploiting the features of user demand, which, in turn, drives us to rethink the optimization process in resource management.

- **Users can play a more active role in improving QoE**. As mentioned above, in most existing standards, network handoff is initiated only when the received signal strength worsens. Today's multimode smartphones and new transmission protocols such as multipath TCP and stream control transmission protocol (SCTP) [22], support fine-grained network selection with low transmission interruption. Users have the motivation to actively adjust network selection when improving QoE is possible. Moreover, users could pay an additional fee to pursue better QoE, for example, paying more fee for priority in sharing radio resource of a cell. Thus, users can balance the cost and QoE optimization.

1.4.2 Motivation of Using Reinforcement Learning

The network selection decision can be quite complex. For the radio access part, the unstable nature of radio channel results in uncertain and dynamic data rate, and the resource sharing among associated users of limited base station bandwidth makes the achieved throughput constrained and dynamic. For the wired network part, potential congestions of certain nodes/parts along the end-to-end transmission path [24] could also be the bottleneck of transmission performance indicating that an end-to-end perspective is needed in some scenarios. In addition, user demand is hard to model and possesses dynamics and heterogeneity.

To tackle this complex decision-making with dynamic and uncertain environment, reinforcement learning (RL) [25, 26] based approaches are introduced to realize intelligent online optimization, which is just the developing trend of future wireless network management [27]. While traditional optimization theory (such as

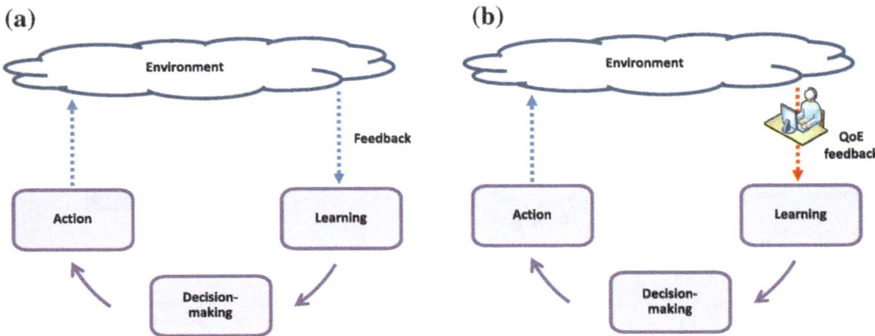

Fig. 1.2 **a** Reinforcement learning loop. **b** User in the reinforcement learning loop in this book

convex optimization) and supervised learning requires prior information or samples working in a relative statistic environment, RL does not rely on prior samples, using a "learning-decision-execution" loop. Note that although RL can also work in offline manner, this book mainly focuses on *online optimization*. The real-time feedback information obtained from the environment is used to asymptotically adjust the decision-making. The advantages of RL online learning are: First, it does not rely on prior environmental information, nor does it need to know the interaction model between the external environment and the decision makers, enabling model-free optimization. Second, it can be used on the fly and takes both the final convergence result and the learning process into account, thus it is suitable for delay-intensive optimization. Finally, it possesses strong robustness by adapting to the external environment dynamics. Different from most existing RL formulations, when RL meets the user-centric optimization in this book, we can achieve users in the learning loop as shown in Fig. 1.2.

The scope of RL is very wide. This book mainly uses two types of RL algorithms. The first type is RL algorithms in sequential decision-making problems in [25]. In a general setting, an agent faces a dynamic environment characterized by states with unknown transition probability and each time the agent selects an action will receive a state-action-specific random reward. The focus is to learn the optimal policy that maps each state to action to maximize the long-term accumulated average reward. Two classical RL learning models, i.e., Markov decision process (MDP) and multi-armed bandit (MAB) problem, are used in network selection in dynamic environment. The second type is multi-agent RL (MARL) in game theory [28]. Game theory is a powerful tool to analyze the decision-making interaction in multiple user systems. Typically, users' network selection problem is formulated as a game model; then the existence and properties of equilibrium points are analyzed to understand the system stability; finally, MARL algorithms are designed to achieve the desired equilibrium in a distributed manner. This book proposes two new game models with promising system efficiency. Note that the multiuser interaction is explicitly modeled in game theory, while it is implicitly embedded in the dynamic environment model in sequential decision-making problems.

Finally, to facilitate fast and flexible online network selection and to deal with possible out of order packets, we assume that transport layer protocols such as ECCP [29] or SCTP [30] supporting UE multi-homing are used.

1.5 Organization and Summary

The main body of this book consists of six chapters, which can be divided into three parts as shown in Fig. 1.3.

- The first part (Chaps. 2 and 3) focuses on how to learn the best network when QoE is revealed beyond QoS under the framework of MAB. Considering the dynamic network state, the network selection problems are formulated as a sequential decision-making problem. Due to the exploring nature of online learning, the application of MAB in network selection will lead to excessive network handoff cost. In the context of MAB modeling, Chap. 2 tries to optimize the update strategy of learning algorithms to constrain the handoff. Chapter 3 focuses on learning the optimal network selection policy while striking a tradeoff between QoE and transmission cost in a dynamic environment. The problem is formulated as a continuous-time MAB and several RL algorithms are proposed.
- The second part (Chaps. 4 and 5) focuses on how to meet dynamic user demand in complex and uncertain heterogeneous wireless networks under the framework of MDP. Chapter 4 focuses on learning the optimal network and traffic type matching policy to strike a tradeoff between QoE and network handoff cost in a dynamic environment. The problem is formulated as a Markov decision process and several Q learning based algorithms are proposed. Since the communication context may change due to the mobility of the user, learning algorithm may need to reset frequently, leading to low learning efficiency. Hence, Chap. 3 explores the idea of transfer learning to boost RL, that is, to use learning experience for identical context.
- The third part (Chaps. 6 and 7) focuses on how to meet heterogeneous user demands of large-scale networks. Game theory is used to analyze the network selection behavior of multiple users. Chapter 6 considers the user-network association problem with heterogeneous user demands in multiple user systems. Due to the heterogeneity in user demand, the complexity of global optimization is high. To this end, two MARL algorithms are designed and theoretically proved to be able to achieve fair balance between optimization complexity and social welfare under a localized cooperation game formulation. Chapter 7 takes a step further to exploit the user demand diversity by proposing a QoE game. The existence and property of QoE equilibrium are analyzed and two MARL algorithms achieving the QoE equilibrium are proposed.

Fig. 1.3 Organization of the main body

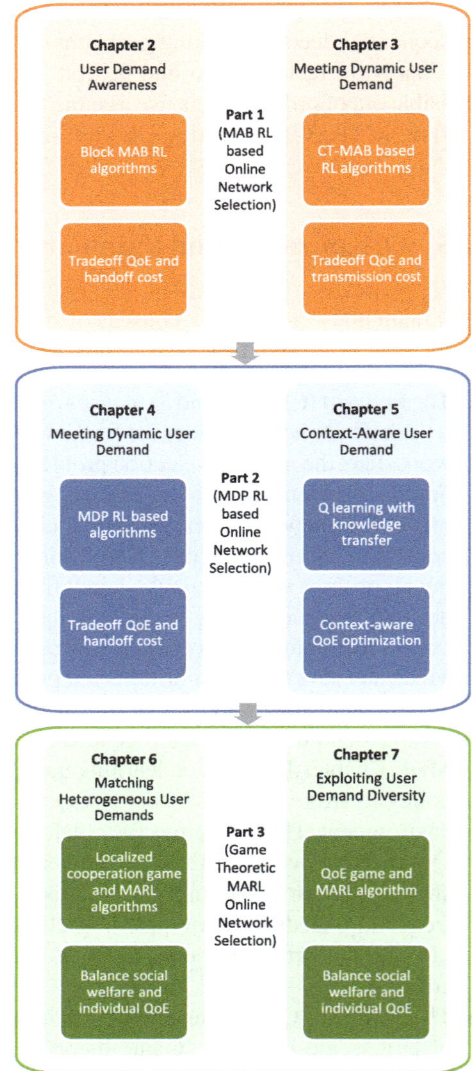

References

1. Zhang N, Cheng N, Gamage AT et al (2015) Cloud assisted HetNets toward 5G wireless networks future of health insurance. IEEE Commun Mag 53:59–65
2. Holland OD, Aijaz A, Kaltenberger F et al (2016) Management architecture for aggregation of heterogeneous systems and spectrum bands. IEEE Commun Mag 54:9–16
3. Du Z, Sun Y, Guo W et al (2018) Data-driven deployment and cooperative self-organization in ultra-dense small cell networks. IEEE Access 6:22839–22848

4. Cisco (2019) Cisco visual networking index: forecast and trends 2017–2022 white paper. Dod J (1999) Effective Substances. In: The dictionary of substances and their effects. Royal Society of Chemistry. https://www.cisco.com/c/en/us/solutions/collateral/service-provider/visual-networking-index-vni/white-paper-c11-741490.html

5. ITU (2017) Opinion model for video-telephony applications. ITU-T Recommendation G.1070 ed

6. Wang PL et al (2017) A data-driven architecture for personalized QoE management in 5G wireless networks. IEEE Wirel Commun 24:102–110

7. Bhushan N, Li J, Malladi D et al (2014) Network densification: the dominant theme for wireless evolution into 5G. IEEE Commun Mag 8(2):82–89

8. Ma D, Ma M (2012) A QoS oriented vertical handoff scheme for WiMAX/WLAN overlay networks. IEEE Trans Mob Comput 23(4):598–606

9. Malanchini I, Cesana M, Gatti N (2013) Network selection and resource allocation games for wireless access networks. IEEE Trans Mob Comput 12(12):2427–2440

10. Nguyen-Vuong Q T, Ghamri-Doudane Y, Agoulmine N (2008) On utility models for access network selection in wireless heterogeneous networks. In: IEEE network operations and management symposium (NOMS)

11. Alexander R, Sebastian M (2011) Recent multimedia QoE standardization activities in ITU-T SG12. IEEE Comsoc Mmtc E-Lett 6(8):10–14

12. Piamrat K, Ksentin A, Viho C (2008) QoE-based network selection for multimedia users in IEEE 802.11 wireless networks. In: IEEE local computer networks (LCN)

13. Hou J, Brien DC (2006) Vertical handover decision making algorithm using fuzzy logic for the integrated radio-and-ow system. IEEE Trans Wirel Commun 5(1):176–185

14. Piamrat K, Ksentin A, Viho C (2008) QoE-aware admission control for multimedia applications in IEEE 802.11 wireless networks. In: IEEE vehicular technology conference (VTC fall)

15. Stevens-Navarro E, Lin Y, Wong VWS (2008) An MDP-based vertical handoff decision algorithm for heterogeneous wireless networks. IEEE Trans Veh Technol 57(2):1243–1254

16. Keshavarz-Haddad A, Aryafar E, Wang M, Chiang M (2017) HetNets selection by clients: convergence, efficiency, and practicality. IEEE ACM Trans Netw 25(1):406–419

17. Zhu K, Hossain E, Niyato D (2014) Pricing, spectrum sharing, and service selection in two-tier small cell networks: a hierarchical dynamic game approach. IEEE Trans Mob Comput 13(8):1843–1856

18. Nguyen DD, Nguyen HX, White LB (2017) Reinforcement learning with network-assisted feedback for heterogeneous RAT selection. IEEE Trans Wirel Commun 16(9):6062–6076

19. Trestian R, Ormond O, Muntean G (2012) Game theory-based network selection: solutions and challenges. IEEE Commun Surv Tutor 14(4):1–20

20. Liu D, Wang L et al (2016) User association in 5G networks: a survey and an outlook. IEEE Commun Surv Tutor 18(2):1018–1044

21. Chen YJ, Wu KS, Zhang Q (2015) From QoS to QoE: a tutorial on video quality assessment. IEEE Commun Surv Tutor 17:1126–1165

22. Wu J, Cheng B, Wang M et al (2017) Quality-aware energy optimization in wireless video communication with multipath TCP. IEEE ACM Trans Netw 25:2701–2718

23. Du Z, Liu D, Yin L (2017) User in the loop: QoE-oriented optimization in communication and networks. In: The 6th international conference on computer science and network technology (ICCSNT)

24. Zhang J, Ansari N (2011) On assuring end-to-end QoE in next generation networks: challenges and a possible solution. IEEE Commun Mag 49:185–191

25. Sutton RS, Barto AG (2017) Reinforcement learning: an introduction, 2nd edn. MIT Press, London

26. Kaelbling LP, Littman ML, Moore AW (1996) Reinforcement learning: a survey. J Artif Intell Res 4:237–285

27. Klaine PV, Imran MA et al (2017) A survey of machine learning techniques applied to self-organizing cellular networks. IEEE Commun Surv Tutor 19(4):2392–2431

28. Xu Y, Wang J, Wu Q et al (2015) A game-theoretic perspective on self-organizing optimization for cognitive small cells. IEEE Commun Mag 53:100–108
29. Arye M et al (2012) A formally-verified migration protocol for mobile, multi-homed hosts. In: IEEE ICNP
30. Wallace T, Shami A (2012) A review of multihoming issues using the stream control transmission protocol. IEEE Commun Surv Tutor 14(2):565–578

Part I

MAB RL Based Online Network Selection

This part studies online network selection considering user demand. Due to uncertain and dynamic network state information, the challenge is how to learn the optimal network. To tackle this challenge, network selection is formulated as a multi-armed bandit (MAB) problem, where the focus is to balance the exploration and exploitation issue during learning process. Since learning process will incur frequent network handoff, Chap. 2 studies MAB with handoff constraint and designs RL algorithms with reduced handoff cost. Chapter 3 further tries to tradeoff transmission cost and QoE under the MAB framework. A continuous-time MAB problem is formulated and three efficient RL algorithms are proposed.

Chapter 2
Learning the Optimal Network with Handoff Constraint: MAB RL Based Network Selection

Abstract The core issue of network selection is to select the optimal network from available network access point (NAP) of heterogeneous wireless networks (HWN). Many previous works evaluate the networks in an ideal environment, i.e., they generally assume that the network state information (NSI) is known and static. However, due to the varying traffic load and radio channel, the NSI could be dynamic and even unavailable for the user in realistic HWN environment, thus most existing network selection algorithms cannot work effectively. Learning-based algorithms can address the problem of uncertain and dynamic NSI, while they commonly need sufficient samples on each option, resulting in unbearable handoff cost. Therefore, this chapter formulates the network selection problem as a multi-armed bandit (MAB) problem and designs two RL-based network selection algorithms with a special consideration on reducing network handoff cost. We prove that the proposed algorithms can achieve optimal order, e.g., logarithmic order regret with limited network handoff cost. Simulation results indicate that the two algorithms can significantly reduce the network handoff cost and improve the transmission performance compared with existing algorithms, simultaneously.

2.1 Introduction

Network selection is of great importance for improving users' communication quality. One of the core issues in network selection is decision-making, i.e., determining the optimal network in terms of transmission performance for users. Traditionally, most existing methods adopted different tools to evaluate and compare the performance of networks according to their NSI, such as received signal strength (RSS), bandwidth, delay, price etc. However, in realistic network environment where no additional protocol or network infrastructure supporting information sharing [1] among HWN, there is no NSI available for users to make decision, which will impose great difficulties on network selection. Alternatively, some network selection or vertical handoff algorithms have been proposed [2, 3] to tackle the dynamic and uncertain issue of NSI. Commonly, the preferred network is learnt by the interaction with the environment.

© Springer Nature Singapore Pte Ltd. 2020
Z. Du et al., *Towards User-Centric Intelligent Network Selection
in 5G Heterogeneous Wireless Networks*,
https://doi.org/10.1007/978-981-15-1120-2_2

However, the dynamic interaction process may lead to excessive network handoff cost. Since the network handoff process includes system discovery, handoff decision, and vertical handoff execution, which will incur additional signaling control cost. Learning algorithms need to explore all options with sufficient samples, resulting in frequent handoff among networks, known as "ping-pong effect". For this reason, the learning process may incur excessive network handoff cost and introduce potential interruptions for service transmission.

In the literature, the dynamic network parameter resulted from the user arrival and departure is formulated as a Markov process and Q-learning is used for maximizing user's QoS in [3]. A weighted bipartite graph based algorithm for multiple flows in heterogeneous wireless network is proposed in [4]. The authors in [2] focus on the resource sharing among multiple users in HWN and Q-learning based algorithms are used to learn the Nash equilibriums. However, the handoff cost is not included in the above works. Some existing works take into account the handoff cost in different ways. The authors in [5] incorporate the handoff cost into the reward and formulate the network selection problem as a Markov decision process. Based on [5], the authors further consider the handoff constraint in [6]. However, they assume that the state transition probabilities on network parameters are known and use the policy iteration algorithm, rather than learning algorithms. In [7, 8], the signaling cost is considered. The handoff latency is treated as the handoff cost in [9]. Note that although these studies are applied learning in network selection, none of them pay attention to the network handoff cost in the learning process. In some works, the handoff cost is an attribute in the utility for making the selection decision, while the long-term handoff cost is uncertain. On the other hand, multi-armed bandit (MAB) based reinforcement learning has been widely studied in communications and networks. The reprehensive algorithms include the index-based policy in [10] that can asymptotically achieve the lower bound of the regret for some families of reward distributions with a single real parameter and classical index-based algorithms that achieve logarithmic order regret for any reward distribution with bounded support uniformly over time in [11], rather than only asymptotically. Some works [12, 13] considered the handoff cost in MAB. However, related algorithms just achieve logarithmic order asymptotically.

Motivated by this fact, we aim to investigate online learning based network selection algorithms with handoff cost consideration. Reducing the network handoff cost in learning–based algorithms is difficult. In online learning theory, the user faces the inherent tradeoff between exploration and exploitation, i.e., keeping enough samplings in suboptimal actions to explore the environment and selecting the potential optimal action as possible. The handoff cost consideration further makes the design of learning algorithms complex, which means that we should carefully balance handoff control and learning process. In this chapter, we formulate the network selection as an online learning problem and propose two learning-based network selection algorithms. Instead of updating learning in each "slot", we reduce the network handoff cost by updating in a block manner. The starting point of this design is that with the progressing of learning, the optimal network is selected with increasing probability, which indicates the expected performance loss of block-based updating algorithms approaches to that of slot-based updating algorithms, and the handoff cost can be significantly reduced. It is interesting to find that compared with some existing algorithms, the proposed algorithms can reduce the handoff cost as well as improve the transmission performance, simultaneously. The main results of this chapter were presented in [14].

2.2 System Model

We consider the NAP set as $\mathcal{N}$, $|\mathcal{N}| = N$. A multimode user equipment (UE) locates in the overlapped coverage area of these N networks. In practical situations, due to the changing traffic load and radio channels, the NSI of NAPs such as bandwidth, delay, etc., are unknown and even dynamic to the user. Note that in this information uncertain situation, most of the existing approaches that rely on known and static NSI cannot work effectively. To tackle this challenge, we will formulate network selection in an online learning framework as shown in Fig. 2.1.

As can be seen in the figure, the system works in a slotted manner. The UE selects a NAP $\delta(t) \in \mathcal{N}$ to transmit at the beginning of slot t. If $\delta(t) \neq \delta(t-1)$, a network handoff happens, otherwise, the UE keeps on transmitting in currently associated NAP. Denote vector $\mathbf{x} = [x_1, x_2, \ldots, x_L]$ as the NSI vector whose elements $x_l, l = 1, 2, \ldots, L$ could be involved L network parameters such as bandwidth, delay, etc. It is assumed that the slot duration is appropriately chosen such that the involved network parameters in $\mathbf{x}$ are approximately constant in a slot, while in different slots, they are independent random variables. At the end of tth slot, the UE get a reward $r(t) = D\left(\mathbf{x}_{\delta(t)}(t)\right)$, where $\mathbf{x}_n(t)$ is the dynamic network parameter vector of network n in the tth slot, the user demand function or QoE

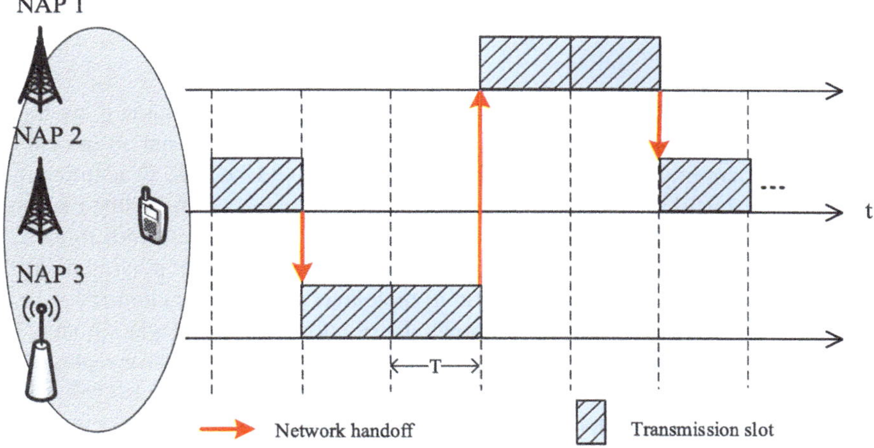

Fig. 2.1 Slotted online network selection framework

function $D(\mathbf{x})$ maps the experienced NSI vector $\mathbf{x}$ to the user's reward such as QoE. We further consider model-free learning framework, where only the end-to-end reward feedback could be used. In other words, both the NSI and utility function are unknown to the user and the selection of $\delta(i)$ is just based on the observation on the access action history $A(t) = \{\delta(1), \delta(2), \ldots, \delta(t-1)\}$ and reward feedback history $U(t) = \{r(1), r(1), \ldots, r(t-1)\}$. A network selection policy π makes the network selection decision for each slot according to the observation $A(t)$ and $U(t)$.

We take the network handoff cost into the model formulation as follows. During a network handoff, the user should first inform the currently associated NAP and disconnect from it. Then, the user must discover and synchronize with the target NAP to finish the new association. Both these two stages incur signaling costs, which depend on specific types of wireless networks, radio technology, protocols, and so on. To account for this feature, we define a handoff cost matrix as $\mathbf{C} = \{c_{m,n}\}, m, n \in \mathcal{N}$, where $c_{m,n}$ is the signaling cost when the user handoffs from NAP m to n. Moreover, the cost is zero when no handoff occurs, i.e., $c_{m,n} = 0$ if $m = n$.

2.3 MAB Problem Formulation

In this section, we formulate the network selection as MAB online learning problem and resort to RL-based algorithms with a special consideration on the network handoff cost.

Due to the dynamics in network parameters, the QoE reward in each slot is random. In this context, the networks' performance is evaluated by the expected rewards $\mu_n = \mathrm{E}[r(\mathbf{x}_n)]$, Naturally, if $\mu_n, n \in \mathcal{N}$ are known, the optimal network n^* for the user is the one with the largest expected reward, i.e.,

$$n^* = \arg\max_{n \in \mathcal{N}} \mu_n. \tag{2.1}$$

When both the prior instant and statistical information on NSI is unknown, we have to learn the optimal network n^*. Then, a key issue arises: How to run a reasonable learning or sampling manner? In other words, we should be able to identify the optimal choice while learning its efficiency during the process. This is just the core of online learning or RL. In the following, we resort to the classical MAB model.

A general MAB problem can be formulated under the scenario of playing N arms of a player. Each time playing some arm n, the player will get a random reward r_n, which follows some arm-specific unknown distribution Φ_n. The player's goal is to find a policy maximizing the long-term expected reward in successive t plays, or equivalently, minimize the regret defined as the gap between the expected reward of the policy and the optimal policy, i.e., always playing the best arm. For the considered network selection problem, the N available networks are the arms of the bandit, the user is the player, accessing one of the NAPs in each slot is equivalent to playing an arm. As a result, the goal is to design a policy maximizing the long-term reward.

Given a network selection policy π, the expected accumulated reward in t slots is defined as

$$E[S_\pi(t)] = E_\pi\left[\sum_{i=1}^{t} r(i)\right]$$

$$= \sum_{n=1}^{N} \mu_n E_\pi[V_n(t)], \tag{2.2}$$

where $V_n(t) = \sum_{i=1}^{t} I\{\delta(i) = n\}$ is the number of slots network n that is selected in the first t slots, $I\{\cdot\}$ is the indicator function and $E_\pi[\cdot]$ means taking expectation under policy π. Meanwhile, the network handoff cost incurred for a network selection policy in the first tth slot is

$$E[H_\pi(t)] = E_\pi\left[\sum_{i=1}^{t} c_{\delta(i-1),\delta(i)}\right] \tag{2.3}$$

To include the network handoff cost in the total reward, we can define the long-term reward is $E[S_\pi(t)] - \phi E[H_\pi(t)]$, where $0 \leq \phi \leq 1$ is the weight for the handoff cost. In our problem, the ideal policy is sticking to the network defined in (2.1) and it incurs no handoff cost. Accordingly, a policy π's regret is defined as

$$E[R_\pi(t)] = E[S_{\pi^*}(t)] - \{E[S_\pi(t)] - \phi E[H_\pi(t)]\}$$

$$= \sum_{n \neq n^*} (\mu_{n^*} - \mu_n) E_\pi[V_n(t)] + \phi E[H_\pi(t)] \tag{2.4}$$

Besides, to avoid frequent network handoff in a limited time, which is known as "ping-pong effect" in network handoff, we define an additional handoff cost metric, rate of handoff cost (RHC), which is the expected handoff cost per slot, as

$$\rho(t) = \frac{E[H_\pi(t)]}{t}. \tag{2.5}$$

With the above-defined two metrics, we are going to find a network selection algorithm or policy π (we will use network selection algorithm and network selection policy interchangeably) to minimize the regret and the RHC

$$\begin{cases} \min_\pi E[R_\pi(t)] \\ \min_\pi \rho(t). \end{cases} \tag{2.6}$$

2.4   Algorithm Design

After formulating the MAB problem, we have to design learning algorithm to achieve the above-defined objective. However, the regret minimization alone is challenging. In MAB, it has been proved that in the optimal policy minimizing the regret is too complex to derive [10], i.e., all the past learning experience is needed and some other conditions should be satisfied. On the contrary, since it has been theoretically proved that the regret grows at least logarithmically with the number of plays, researchers mainly focus on design low-complexity order-optimal policies [11]. Hence, we also adhere to this criterion. On this basis, we try to control the number of switching arms in the learning algorithm:

1. Achieving logarithmic order regret $E[R_\pi(t)]$;
2. Achieving controlled RHA.

There exist several MAB RL algorithms that achieve the logarithmic order regret, while limited work explicitly focused on the accumulative switching cost and handoff frequency. In most existing learning algorithms, the arm selection decision is made for each play (slot), while our idea is reducing network handoff cost by maintaining each decision for multiple successive plays, which is called block-based RL here. Specifically, we envision two types of block patterns: one is constant block length, the other one is variable block length. For constant block length, we can keep each action decision for constant successive plays. For variable block length, we control the block length to gradually grow with time. A question arises that how to set the block pattern to reduce the switching cost without violating the logarithmic regret property? In the following, we present and validate two algorithms, block UCB1 based algorithm and UCB2 based algorithm, respectively.

2.4.1   Block UCB1 Algorithm Based Algorithm

The first idea is generalizing the updating manner of classical MAB learning algorithm to reduce arm switch cost. We construct the constant block length MAB RL algorithm based on the classical UCB1 [11], which is a widely used online learning algorithm in MAB. We modify the UCB1 to get a block UCB1 based network selection algorithm as shown in Algorithm 1. The proposed algorithm works as follows: At the beginning, each NAP is selected for one block, i.e., m successive slots, where m is an integer larger than one. After that, the network maximizing $\hat{r}_n + \sqrt{\frac{2m \log k}{\gamma_n}}$ is selected for one block, k is the block index, $\hat{r}_n$ is the sample mean of the reward for the transmission in network n, γ_n is the number of blocks network n is selected.

The performance of block UCB1 algorithm is provided by Theorem 6.1.

Theorem 2.1 *The regret of block UCB1 algorithm is order-optimal, i.e.*

Algorithm 1 Block UCB1 Algorithm

1: **Initiate:** Block length m, number of networks N.
2: **loop**
3:   **if** $k \leq N$ **then**
4:     Select $\delta(k) = k$ for one block of m slots.
5:   **else**
6:     Select $\delta(k) = \arg\max_{n \in \mathcal{N}} \left\{ \hat{r}_n + \sqrt{\frac{2m \log k}{\gamma_n}} \right\}$ for successive m slots.
7:   **end if**
8:   Update $\gamma_{\delta(k)} = \gamma_{\delta(k)} + 1$, and $\hat{r}_{\delta(k)} = \dfrac{\sum\limits_{g=1}^{k+1} \sum\limits_{t=(g-1)m-1}^{gm} I\{\delta(t)=\delta(k)\} r(t)}{\gamma_{\delta(k)}}$
9: **end loop**

$$E[R_{block}(t)]$$
$$\leq \sum_{n:\mu_n < \mu_{n*}} \left(m\,\Delta_n + \phi c'_{n*} + \phi c'_n \right) \left[1 + \frac{8 \log t}{m \Delta_n^2} + \frac{\pi^2}{3} \right]. \tag{2.7}$$

Its RHC is upper bounded as

$$\rho_{block}(t)$$
$$\leq \sum_{n:\mu_n < \mu_{n*}} \frac{\left(c'_{n*} + c'_n \right)}{t} \left[1 + \frac{8 \log t}{m \Delta_n^2} + \frac{\pi^2}{3} \right], \tag{2.8}$$

where $\Delta_n = \mu_{n*} - \mu_n$, $c'_n = \max\limits_{m} c_{m,n}$ is the maximal handoff cost to the optimal network n.

We can see from Theorem 2.1 that the block UCB1 algorithm not only achieves the logarithmic-order regret, but also controlled RHC. Moreover, m is an adjustable parameter for both the regret and RHC. In essence, the block UCB1 algorithm is just a generalization of UCB1.

2.4.2 UCB2 Algorithm Based Algorithm

Different from the constant block length, another similar approach is using variable length block for algorithm update. Fortunately, the UCB2 learning algorithm in [11] provides a reference. Although the original UCB2 aims at the order-optimal regret without the handoff cost consideration, we found it works in a block similar to the way with variable block length, which may contribute to reducing switching cost. We then directly use the UCB2-based network selection algorithm as shown in Algorithm 2. The algorithm works as follows: At the beginning, the first N slots are transmitted in each NAP once to ensure that all the NAPs are accessed by the user. From then on, the algorithm is updated with an increasing length of blocks. At the end of each block, the network n' for the next block is determined by

$$n' = \arg\max_{n \in \mathcal{N}} \left\{ \hat{r}_n + a_{i,\tau_n} \right\}, \tag{2.9}$$

where $\hat{r}_n$ is the sample mean of the reward by transmitting a slot in network n, γ_n is the number of blocks in which network n is selected, i is the total block index, $v(\gamma) = \lceil (1+\alpha)^\gamma \rceil$, $a_{i,\gamma} = \sqrt{\frac{(1+\alpha)\log[ei/v(\gamma)]}{2v(\gamma)}}$, $0 < \alpha < 1$ is a parameter controlling the growth pattern of the block length. Note that $v\left(\gamma_{\delta(i)} + 1\right) - v\left(\gamma_{\delta(i)}\right)$ is the block length that generally increases with $\gamma_{\delta(i)}$, that is, the more a network is selected, the block length in which this network is selected is larger.

Algorithm 2 UCB2-based Algorithm

1: **Initiate:** $\gamma_n = 0, n \in \mathcal{N}, i = 0$.
2: **loop**
3:    **if** $i \leq N$ **then**
4:        Select network $\delta(i) = i$ for one slot.
5:    **else**
6:        Select $\delta(i) = \arg\max_{n \in \mathcal{N}} \left\{ \hat{r}_n + a_{i,\gamma_n} \right\}$ for successive $v\left(\gamma_{\delta(i)} + 1\right) - v\left(\gamma_{\delta(i)}\right)$ slots.
7:    **end if**
8:    Update $\gamma_{\delta(i)} = \gamma_{\delta(i)} + 1$, $\hat{r}_n$ is the sample mean of reward for transmitting a slot in network n.
9: **end loop**

Based on the original regret of UCB2 algorithm, which is equivalent to $\sum_{n \neq n^*} (\mu^* - \mu_n) E_\pi [V_n(t)]$, we are able to obtain the RHC and regret bounds of UCB2 based algorithm in Theorem 6.2.

Theorem 2.2 *The RHC of UCB2 based network selection algorithm for any* $t \geq \max_{n:\mu_n < \mu_{n^*}} \frac{1}{2\Delta_n^2}$ *is upper bounded as*

$$\rho_{UCB2}(t) \leq \sum_{n:\mu_n < \mu_{n^*}} \frac{\left(c'_n + c'_{n^*}\right)}{t} \kappa_{n,t}, \tag{2.10}$$

$$\kappa_{n,t} = 1 + \frac{1}{\log(1+\alpha)} \log \left[\frac{(1+4\alpha)\log\left(2et\Delta_n^2\right)}{2\Delta_n^2} \right] +$$

$$\log \frac{\log t}{\log(1+\alpha)} \left\{ \exp \left[-\frac{\alpha^2(1+\alpha)}{4} \right] + \exp(-1-\alpha) \right.$$

$$\left. + \frac{11(1+\alpha)\exp(-1-\alpha)}{5\alpha^2\Delta_n^2\log(1+\alpha)} \right\}, \tag{2.11}$$

its regret is upper bounded as

$$E\left[R_{UCB2}\left(t\right)\right] \leq \phi \sum_{n:\mu_n<\mu_{n*}} \left(c'_n + c'_{n*}\right) \kappa_{n,t}$$

$$+ \sum_{n:\mu_n<\mu_{n*}} \frac{(1+\alpha)\,(1+4\alpha)\log\left(2et\,\Delta_n^2\right) + 2c_\alpha}{2\Delta_n} \tag{2.12}$$

where $c'_n = \max\limits_{k\in N} c_{k,n}$ *is the maximal handoff cost to network* n, $\Delta_n = \mu_{n*} - \mu_n$.

As can be seen that the RHC bound is in the form of $\frac{\log(\log t)}{t}$, which eventually decreases as t increases. Compared with block UCB1 based algorithm, it seems that the RHC in UCB2 algorithm has a faster decreasing speed. α is the key parameter for both the regret and RHC.

2.5 Simulation Results

Some simulation results are presented in this section.

2.5.1 Simulation Setting

We assume that a user located in the overlapping area of three networks, i.e., WLAN1 + WLAN2 + LTE, run the proposed two RL algorithms. Following with [5], we use the discrete model to represent the bandwidth and delay dynamics in networks, where the bandwidth and delay in networks are approximated to be several levels. For the three networks, the unit bandwidth b_{unit} and unit delay d_{unit} are [0.3 Mbps, 50 ms], [0.3 Mbps, 50 ms], and [0.25 Mbps, 50 ms], respectively. The maximal bandwidth B_{max}, minimal bandwidth B_{min}, maximal delay D_{max}, and minimal delay D_{min} are [$10b_{unit}$, $3b_{unit}$, $6d_{unit}$, $2d_{unit}$], [$8b_{unit}$, $3b_{unit}$, $10d_{unit}$, $2d_{unit}$], and [$8b_{unit}$, $4b_{unit}$, $6d_{unit}$, $4d_{unit}$], respectively. In each slot, the joint bandwidth-delay state in each network can be one of all possible combinations with the same probability. The network is stationary if the statistical distribution of the bandwidth-delay state remains unchanged in different slots.

Since WLAN1 and WLAN2 have similar radio technology, the handoff costs between them are relatively smaller. On the other hand, LTE and WLAN have totally different radio technologies, the corresponding handoff costs are relatively larger. In the simulation, the nominal handoff costs among these networks are defined as: 0.2 between WLAN1 and WLAN2, 0.3 between WLAN1 and LTE, 0.5 between WLAN2 and LTE.

The utility function can incorporate both technical aspect and nontechnical aspect [15, 16]. Technical aspect includes QoS parameters such as delay, throughput and packet loss rate. Nontechnical aspect may include economical parameters, user preference, etc. Following the same way with [5], we mainly focus on the impact of the bandwidth and the delay on user's reward and define the utility function $r(b, d)$ as

$$r(b, d) = w f_1(b) + (1 - w) f_2(d), \tag{2.13}$$

where b and d are the experienced bandwidth and delay, respectively. $f_1(b)$ is a bandwidth-related reward and $f_2(d)$ is a delay-related reward,

$$f_1(b) = \begin{cases} 1, & b \geq b_{\max} \\ (b - b_{\min}) / (b_{\max} - b_{\min}), & b_{\min} < b < b_{\max} \\ 0, & b \leq b_{\min} \end{cases} \tag{2.14}$$

$$f_2(d) = \begin{cases} 1, & 0 < d \leq d_{\min} \\ (d_{\max} - d) / (d_{\max} - d_{\min}), & d_{\min} < d < d_{\max} \\ 0, & d \geq d_{\max}. \end{cases} \tag{2.15}$$

$0 \leq w \leq 1$ is the weight determining the importance of the bandwidth reward and the delay reward. In $f_1(b)$, $b_{\min}$ and $b_{\max}$ are two bandwidth thresholds representing the required minimal and maximal bandwidths, respectively. When the achieved bandwidth is larger than the maximal bandwidth $b_{\max}$, the user cannot benefit any more. On the other hand, the minimal bandwidth $b_{\min}$ is the bandwidth threshold to obtain positive reward. The bandwidth reward increases linearly with the achieved bandwidth when it is in the range of $[b_{\min}, b_{\max}]$. Similarly, $d_{\min}$ and $d_{\max}$ are the two delay thresholds in $f_2(d)$. The bandwidth thresholds, delay thresholds, and w can be set either according to some recommendations [17] or user preferences. In the simulation, $b_{\min} = 0.5\,\text{Mbps}$, $b_{\max} = 8\,\text{Mbps}$, $d_{\min} = 100\,\text{ms}$, $d_{\max} = 600\,\text{ms}$, $w = 0.5$.

2.5.2 Results

Figures 2.2 and 2.3 show the sample runs of block UCB1 based algorithm ($m = 10$) and UCB2 based algorithm ($\alpha = 0.05$). We can observe that the selection ratio of the best NAP (WLAN1) is dominant and it increases as time in both algorithms. However, the evolution patterns of the two algorithms are different due to different update rules and block lengths. In particular, the linear growth episodes indicate that the block length in block UCB1 based algorithm is kept constant and increases in UCB2-based algorithm.

We adjust the two parameters m and α to test their impacts on the algorithm performance. The regrets of block UCB1 based network selection algorithm with

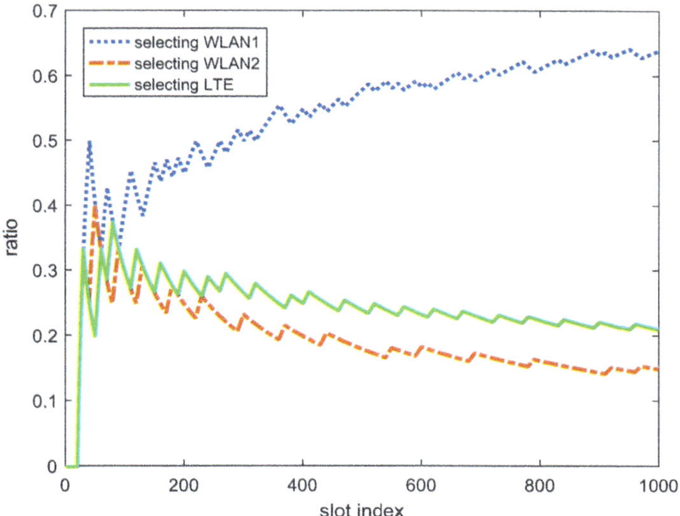

Fig. 2.2   A sample run of block UCB1 based algorithm

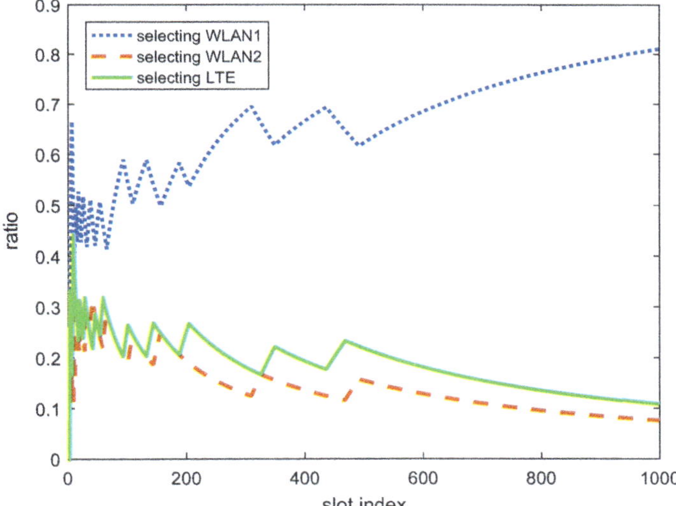

Fig. 2.3   A sample run of UCB2 based algorithm

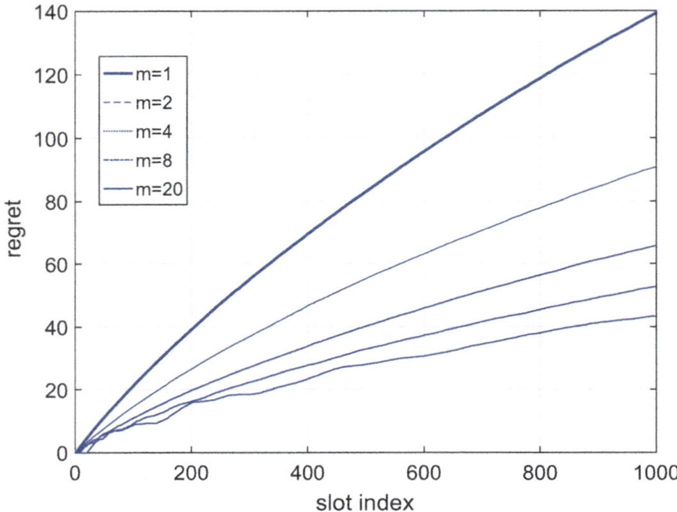

Fig. 2.4 The regret of block UCB1 based algorithm

different m averaged by 500 runs with 1000 slots are presented in Fig. 2.4. It is seen that all the regrets grow in a sub-linear manner with the slot index. Compared with UCB1, the block UCB1 based algorithms with $m > 1$ achieve significant smaller regret. Especially, a larger m results in a smaller regret. The results validate both the logarithmic order regret and the superiority of the block UCB1 algorithm. However, the fluctuation of the regret becomes more serious when m is larger than 8. This is because that when m is too large, there exist short episodes with linear growth in regret. For this reason, m should not be too large in practice. The RHC of UCB1 based algorithms in Fig. 2.5 show that the RHC appears a peak in the first 100 slots and gradually decreases with slot index for all five cases. This can be explained as that in the early stage of learning algorithms, frequent network handoff occurs for exploring the environment. The optimal network is selected more and more often as the learning algorithm progresses, thus, handoff becomes more and more rare. Generally, the peak RHC as well as the stable RHC decrease as the m increases and the case of $m = 1$ incurs the largest RHC. These results indicate that the handoff cost reduction does not necessarily lead to reward loss. In case of the short-term linear growth in the regret, a median m such as $1 < m < 20$ is recommended.

Figures 2.6 and 2.7 show the regret and RHC of UCB2 based network selection algorithm with different α. As expected, for a given α, the regret grows sub-linearly with the slot index and a larger α generally leads to a smaller regret. Also, the regrets show fluctuation when α exceeds some threshold, e.g. $\alpha \geq 0.4$. Similar to block UCB1 based algorithm, the RHC of UCB2 based algorithm has a peak in the first 100 slots and then decreases. Different form block UCB1 based algorithm, the peak

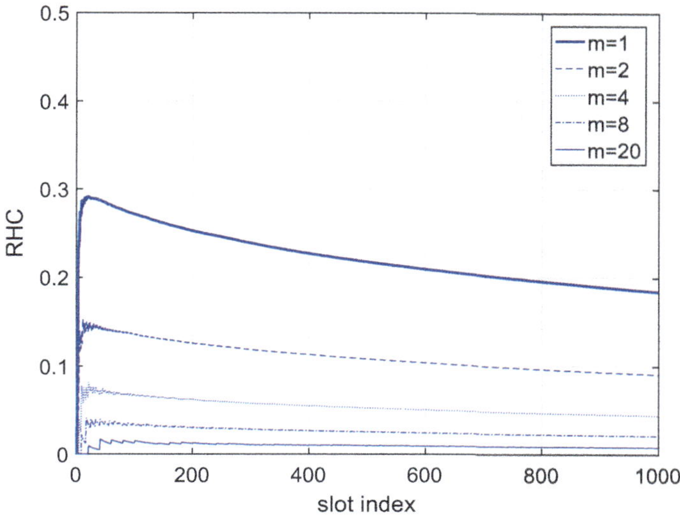

Fig. 2.5 The RHC of block UCB1 based algorithm

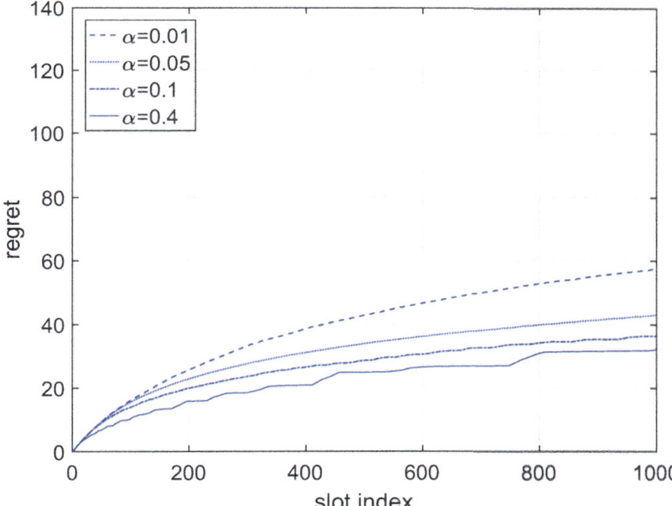

Fig. 2.6 The regret of UCB2 based algorithm

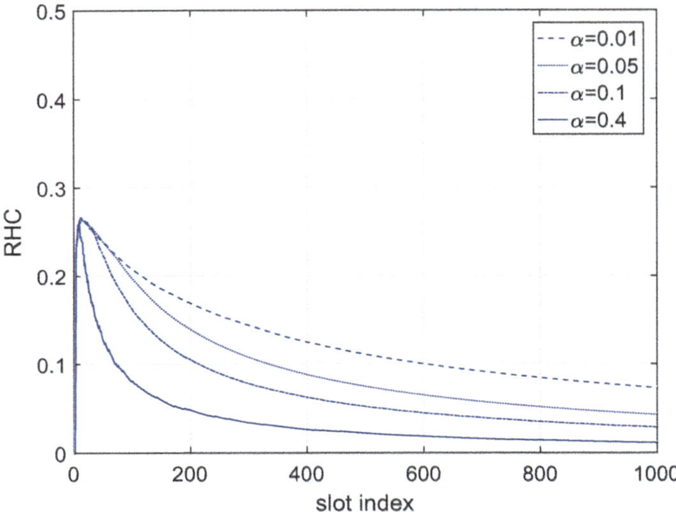

Fig. 2.7 The RHC of UCB2 based algorithm

RHC almost keeps unchanged for different α. This is due to the fact that the block length is small at the early stage of UCB2 based algorithm independent with α. We can conclude that both the regret and RHC of UCB2 based algorithm decreases as α increase, according to (7) and (9). $0.01 \leq \alpha \leq 0.4$ is feasible in this context.

The above results demonstrated the effectiveness of the proposed two algorithms. In addition, as key parameters affect the algorithm performance, appropriate parameter settings is important for the algorithms.

2.6 Conclusion

In this chapter, the network selection problem with dynamic NSI is formulated as a MAB problem. The goal is to efficiently learn the optimal network with controlled network handoff cost. Two RL-based network selection algorithms block UCB1 and UCB2 based algorithms are proposed with a special consideration on reducing the handoff cost. The main idea is adopting block-based updating manner to reduce network handoff number. The regrets of the proposed algorithms are proved to be order-optimal and the handoff cost is analyzed to be bounded. Finally, simulation results verified the algorithms and analyzed their properties.

Appendix

Appendix 1: Proof of Theorem 6.1

Proof Given the slot index t, the block number $K = \lceil \frac{t}{m} \rceil$ in block UCB1 based algorithm. $\gamma_n(K)$ is the number of blocks that network n is selected in the first K blocks

$$
\begin{aligned}
\gamma_n(K) &= 1 + \sum_{k=N+1}^{K} I\{\delta(km) = n\} \\
&\leq l + \sum_{k=N+1}^{K} I\{\delta(km) = n, \gamma_n(k-1) \geq l\} \\
&\leq l + \sum_{k=N+1}^{K} I\left\{\hat{r}_{n^*}(k) + \sqrt{\frac{2m \log k}{\gamma_{n^*}(k)}} \leq \hat{r}_n(k) + \sqrt{\frac{2m \log k}{\gamma_n(k)}}, \gamma_n(k-1) \geq l\right\} \\
&= l + \sum_{k=N+1}^{K} I\left\{\frac{\hat{r}_{n^*}(k)}{m} + \sqrt{\frac{2 \log k}{m \gamma_{n^*}(k)}} \leq \frac{\hat{r}_n(k)}{m} + \sqrt{\frac{2 \log k}{m \gamma_n(k)}}, \gamma_n(k-1) \geq l\right\}
\end{aligned}
$$

Note that $\frac{\hat{r}_n(k)}{m}$ is the sample mean of r_n, based on the Chernoff–Hoeffding bound [11], we can get that

$$
\begin{aligned}
E\left[\gamma_n(K)\right] &\leq \frac{8 \log K}{m \Delta_n^2} + 1 + \frac{\pi^2}{3} \\
&\leq \frac{8 \log\left(\frac{t+m}{m}\right)}{m \Delta_n^2} + 1 + \frac{\pi^2}{3} \\
&= \frac{8 \log(t+m) - 8 \log m}{m \Delta_n^2} + 1 + \frac{\pi^2}{3} \\
&< \frac{8 \log t}{m \Delta_n^2} + 1 + \frac{\pi^2}{3}
\end{aligned}
\tag{2.16}
$$

The handoff cost bound is

$$
\begin{aligned}
E\left[H_{block}(t)\right] &\leq E\left[\sum_{n: \mu_n < \mu_{n^*}} c'_n \gamma_n(K)\right] + c'_{n^*} E\left[\sum_{n: \mu_n < \mu_{n^*}} \gamma_n(K)\right] \\
&= \sum_{n: \mu_n < \mu_{n^*}} c'_n E\left[\gamma_n(K)\right] + c'_{n^*} \sum_{n: \mu_n < \mu_{n^*}} E\left[\gamma_n(K)\right] \\
&\leq \sum_{n: \mu_n < \mu_{n^*}} \left(c'_{n^*} + c'_n\right) \left[\frac{8 \log t}{m \Delta_n^2} + 1 + \frac{\pi^2}{3}\right]
\end{aligned}
\tag{2.17}
$$

where $\Delta_n = \mu_{n^*} - \mu_n$, $c'_{n^*} = \max\limits_{m} c_{m,n^*}$ is the maximal handoff cost for a handoff to the optimal network n^*, $K = \left\lceil \frac{t}{m} \right\rceil$ is the block number for t. Note that the above case corresponds to the worst scenario, where any two suboptimal network selections are separated by one optimal network selection.

The expected regret can be bounded as

$$E\left[R_{block}\left(t\right)\right] \leq \sum_{n:\mu_n<\mu_{n^*}} m\Delta_n E\left[\gamma_n\left(K\right)\right] + \phi E\left[H_\pi\left(t\right)\right]$$

$$\leq \sum_{n:\mu_n<\mu_{n^*}} \left(m\Delta_n \phi c'_{n^*} + \phi c'_n\right) \left[\frac{8\log t}{m\Delta_n^2} + 1 + \frac{\pi^2}{3}\right] \tag{2.18}$$

This completes the proof. $\square$

Appendix 2: Proof of Theorem 2.2

Proof Similarly, the handoff cost and regret for UCB2 based algorithm are as follows:

$$E\left[H_{UCB2}\left(t\right)\right] \leq E\left[\sum_{n:\mu_n<\mu_{n^*}} c'_n\gamma_n\right] + c'_{n^*} E\left[\sum_{n:\mu_n<\mu_{n^*}} \gamma_n\right]$$

$$= \sum_{n:\mu_n<\mu_{n^*}} \left(c' + c'_{n^*}\right) E\left[\gamma_n\right] \tag{2.19}$$

$$E\left[R_{UCB2}\left(t\right)\right] \leq \sum_{n:\mu_n<\mu_{n^*}} \left(\mu^* - \mu_n\right) E\left[\gamma_n\right] + \phi E\left[H_{UCB2}\left(t\right),\right] \tag{2.20}$$

where $c'_n = \max\limits_{k\in N} c_{k,n}$ is the maximal handoff cost when the selected network is n. We bound $H_{UCB2}\left(t\right)$ and $R_{UCB2}\left(t\right)$ by finding the bound for $E\left[\gamma_n\right]$, $n \neq n^*$, which is the expected number of blocks in which network n is selected. For any $t \geq \max\limits_{\mu_n<\mu_{n^*}} \frac{1}{2\Delta_n^2}$, $\Delta_n = \mu_{n^*} - \mu_n$, denote $\tilde{\gamma}_n$ as the largest integer satisfying

$$v\left(\tilde{\gamma}_n - 1\right) \leq \frac{\left(1 + 4\alpha\right)\log\left(2et\Delta_n^2\right)}{2\Delta_n^2}$$

According to the definition of $v\left(\tau\right)$, $v\left(\gamma_n\right) \leq t$ for $\forall n$, then

$$\gamma_n \leq \frac{\log t}{\log\left(1 + \alpha\right)}.$$

Therefore,

$$\gamma_n \leq 1 + \sum_{\gamma > 1}^{\frac{\log t}{\log(1+\alpha)}} I\{\text{network } n \text{ has started } \gamma\text{th block}\}$$

$$\leq \tilde{\gamma}_n + \sum_{\gamma > \tilde{\gamma}_n}^{\frac{\log t}{\log(1+\alpha)}} I\{\text{network } n \text{ has started } \gamma\text{th block}\}. \tag{2.21}$$

According to [11], the event "network n has started γ-th block" implies $\exists i \geq 0$ such that at least one of the following two events holds:

$$\hat{r}_{n,v(\gamma-1)} + a_{t,\gamma-1} \geq \mu^* - \frac{\alpha \Delta_n}{2}$$

$$\hat{r}^*_{v(i)} + a_{v(\gamma-1)+v(i),i} \leq \mu^* - \frac{\alpha \Delta_n}{2}$$

Thus,

$$E\left[\gamma_n\right] \leq \tilde{\gamma}_n + \sum_{\gamma > \tilde{\gamma}_n}^{\frac{\log t}{\log(1+\alpha)}} P\{\hat{r}_{n,v(\gamma-1)} + a_{t,\gamma-1} \geq \mu_{n^*} - \tfrac{\alpha \Delta_n}{2}\}$$

$$+ \sum_{\gamma > \tilde{\gamma}_n}^{\frac{\log t}{\log(1+\alpha)}} \sum_{i \geq 0} P\{\hat{r}^*_{v(i)} + a_{v(\gamma-1)+v(i),i} \leq \mu_{n^*} - \tfrac{\alpha \Delta_n}{2}\}$$

In the following, we bound $E\left[\gamma_n\right]$ by bounding its components. Since $v\left(\tilde{\gamma}_n - 1\right) \leq \frac{(1+4\alpha)\log\left(2et\Delta_n^2\right)}{2\Delta_n^2}$, $(1+\alpha)^{\tilde{\gamma}_n - 1} \leq \left\lceil (1+\alpha)^{\tilde{\gamma}_n - 1} \right\rceil = v\left(\tilde{\gamma}_n - 1\right)$, then,

$$\tilde{\gamma}_n \leq 1 + \frac{1}{\log\left(1+\alpha\right)} \log\left[\frac{(1+4\alpha)\log\left(2et\Delta_n^2\right)}{2\Delta_n^2}\right]. \tag{2.22}$$

Denote

$$A\left(n\right) = P\{\hat{r}_{n,v(\gamma-1)} + a_{t,\gamma-1} \geq \mu_{n^*} - \frac{\alpha \Delta_n}{2}\},$$

$$B = \sum_{\gamma > \tilde{\gamma}_n}^{\frac{\log t}{\log(1+\alpha)}} \sum_{i \geq 0} P\{\hat{r}^*_{v(i)} + a_{v(\gamma-1)+v(i),i} \leq \mu_{n^*} - \frac{\alpha \Delta_n}{2}\}$$

According to [11],

$$A\left(n\right) = P\{\hat{r}_{n,v(\gamma-1)} + a_{t,\gamma-1} \geq \mu_n + \Delta_n - \frac{\alpha\Delta_n}{2}\}$$

$$\leq \exp\left\{-v\left(\gamma-1\right)\Delta_n^2\alpha^2/2\right\} \tag{2.23}$$

In addition, $t \geq \max_{\mu_n < \mu_{n*}} \frac{1}{2\Delta_n^2}$, thus,

$$v\left(\gamma-1\right) > \frac{(1+4\alpha)\log\left(2et\Delta_n^2\right)}{2\Delta_n^2}$$

$$\geq \frac{(1+4\alpha)}{2\Delta_n^2}. \tag{2.24}$$

At last, using Chernoff–Hoeffding bound,

$$\begin{aligned}
B &= \sum_{\gamma > \tilde{\gamma}_n}^{\frac{\log t}{\log(1+\alpha)}} \sum_{i \geq 0} P\{\hat{r}^*_{v(i)} + a_{v(\gamma-1)+v(i),i} \leq \mu^* - \tfrac{\alpha\Delta_n}{2}\} \\
&\leq \sum_{\gamma > \tilde{\gamma}_n}^{\frac{\log t}{\log(1+\alpha)}} \sum_{i \geq 0} \exp\left\{-\frac{v(i)\alpha^2\Delta_n^2}{2} - (1+\alpha)\log\left[e^{\frac{v(\gamma-1)+v(i)}{v(i)}}\right]\right\} \\
&\leq \sum_{i \geq 0} v\left(i\right)\exp\left\{-\frac{v(i)\alpha^2\Delta_n^2}{2}\right\} \sum_{\gamma > \tilde{\gamma}_n}^{\frac{\log t}{\log(1+\alpha)}} \exp\left(-1-\alpha\right)\left[\frac{v(i)}{v(\gamma-1)+v(i)}\right]^{1+\alpha} \\
&\leq \sum_{i \geq 0} v\left(i\right)\exp\left\{-\frac{v(i)\alpha^2\Delta_n^2}{2}\right\} \exp\left(-1-\alpha\right)\frac{\log t}{\log(1+\alpha)}.
\end{aligned}$$

From [11],

$$\sum_{i \geq 0} v\left(i\right)\exp\left\{-\frac{v\left(i\right)\alpha^2\Delta_n^2}{2}\right\} < 1 + \frac{11\left(1+\alpha\right)}{5\alpha^2\Delta_n^2\log\left(1+\alpha\right)}$$

. Then,

$$B \leq \left[1 + \frac{11\left(1+\alpha\right)}{5\alpha^2\Delta_n^2\log\left(1+\alpha\right)}\right]\frac{e^{-(1+\alpha)}\log t}{\log\left(1+\alpha\right)}. \tag{2.25}$$

Combining (2.22), (2.24), and (2.25), we get

$$\begin{aligned}
E\left[\gamma_n\right] &\leq 1 + \frac{1}{\log(1+\alpha)}\log\left[\frac{(1+4\alpha)\log\left(2et\Delta_n^2\right)}{2\Delta_n^2}\right] + \\
&\quad \frac{\log t}{\log(1+\alpha)}\left\{\exp\left[-\frac{\alpha^2(1+\alpha)}{4}\right] + \exp\left(-1-\alpha\right) + \frac{11(1+\alpha)\exp(-1-\alpha)}{5\alpha^2\Delta_n^2\log(1+\alpha)}\right\},
\end{aligned}$$

which completes the proof. $\qquad\qquad\square$

References

1. Fernandes S, Karmouch A (2012) Vertical mobility management architectures in wireless networks: a comprehensive survey and future directions. IEEE Commun Surv Tutor 14(1):45–63
2. Niyato D, Hossain E (2009) Dynamics of network selection in heterogeneous wireless networks: an evolutionary game approach. IEEE Trans Veh Technol 58(4):2008–2017
3. Tabrizi H, Farhadi G, Cioffi J (2011) A learning-based network selection method in heterogeneous wireless systems. In: IEEE global telecommunications conference (GLOBECOM)
4. Zhang Y, Yuan Y, Zhou J et al (2009) A weighted bipartite graph based network selection scheme for multi-flows in heterogeneous wireless network. In: IEEE global telecommunications conference (GLOBECOM)
5. Stevens-Navarro E, Lin Y, Wong VWS (2008) An MDP-based vertical handoff decision algorithm for heterogeneous wireless networks. IEEE Trans Veh Technol 57(2):2008–2017
6. Stevens-Navarro E, Wong VWS (2008) A constrained MDP-based vertical handoff decision algorithm for 4G wireless networks. In: IEEE international conference on communications (ICC)
7. Wu T, Jing H, Yu X et al (2008) Cost-aware handover decision algorithm for cooperative cellular relaying networks. In: IEEE vehicle technology conference (VTC)
8. Wang L, Binet D (2009) Best permutation: a novel network selection scheme in heterogeneous wireless networks. In: International conference on wireless communications and mobile computing (IWCMC)
9. Hou J, O'Brien DC (2006) Vertical handover decision making algorithm using fuzzy logic for the integrated radio-and-OW system. IEEE Trans Wirel Commun 5(1):176–185
10. Lai T, Robbins H (1985) Asymptotically efficient adaptive allocation rules. Adv Appl Math 6:4–22
11. Auer P, Cesa-Bianchi N, Fischer P (2002) Finite-time analysis of the multiarmed bandit problem. Mach Learn 47:235–256
12. Agrawal R, Teneketzis D, Anantharam V (1998) Asymptotically efficient adaptive allocation rules for the multiarmed bandit problem with switching. IEEE Trans Autom Control 33(10):899–906
13. Chen L, Iellamo S, Coupechoux M (2011) Opportunistic spectrum access with channel switching cost for cognitive radio networks. In: IEEE international conference on communications (ICC)
14. Du Z, Wu Q, Yang P (2016) Learning with handoff cost constraint for network selection in heterogeneous wireless network. Wirel Commun Mob Comput 16(4):441–458
15. Zhao T, Liu Q, Chen CW (2017) QoE in video transmission: a user experience-driven strategy. IEEE Commun Surv Tutor 19(1):285–302
16. Quoc-Thinh N, Agoulmine N, Cherkaoui EH et al (2016) Multicriteria optimization of access selection to improve the quality of experience in heterogeneous wireless access networks. IEEE Trans Veh Technol 62(4):1785–1800
17. ITU-T (2003) One-way transmission time. Rec. G.114

Chapter 3
Meeting Dynamic User Demand with Transmission Cost Awareness: CT-MAB RL Based Network Selection

Abstract The access and transmission in wireless networks not only satisfy user demand for service delivery, but also incur cost in terms of fee and energy consumption, which poses the concern of cost-performance ratio. This chapter studies the cost-performance ratio optimization with dynamic traffic types in dynamic and uncertain HWN. To balance the QoE reward and transmission cost for different traffic types on the fly, the problem is formulated as a continuous-time multi-armed bandit (CT-MAB) model. A traffic-aware online network selection algorithm (ONES) is designed to match typical traffic types (user demand) with respective optimal networks in terms of QoE. In addition, we exploit the correlation feature among multiple traffic types to improve the learning capability, which inspires us to propose another two efficient algorithms: decoupled online network selection algorithm (D-ONES) and virtual multiplexing ONES (VM-ONES). Simulation results demonstrate that our online network selection algorithms achieve better QoE reward rate over non-learning-based algorithms and learning-based algorithms without QoE considerations.

3.1 Introduction

Similar to Chap. 2, this chapter focuses on the network selection issue when NSI is dynamic and uncertain. In addition, more challenging and practical considerations are involved: transmission cost and traffic type dynamics. It is well known that accessing either access points in WLAN or base stations in cellular networks often incurs costs such as energy consumption and transmission fee. As a result, the network providing a user with the best QoE may not be preferred if higher energy consumption is needed or the user has to pay more fee. In other words, we need to optimize the cost–performance ratio, i.e., the network transmission cost and QoE reward. In particular, this balance is to be optimized online due to dynamic and uncertain NSI. Moreover, smartphone nowadays supports various traffic types such as high-definition video, voice or video call, and file downloading, thus, it is reasonable to consider traffic type dynamics. Since matching one traffic type with its optimal network corresponds

© Springer Nature Singapore Pte Ltd. 2020
Z. Du et al., *Towards User-Centric Intelligent Network Selection
in 5G Heterogeneous Wireless Networks*,
https://doi.org/10.1007/978-981-15-1120-2_3

to one learning task, the dynamic traffic case faces multiple learning tasks. Hence, how to improve learning efficiency is a new challenge.

In the literature, a variety of influencing factors including network metrics, device-related criteria, traffic requirements, and user preferences [1, 2] can be considered to evaluate the overall network performance. Accordingly, the multiple attribute decision-making (MADM) [3] is widely used as the evaluation method. Note that these related works could consider transmission cost as a factor, but did not explicitly introduce the ratio between QoE reward and transmission cost. On the other hand, although QoE-based network selection has been studied in [4, 5], they neither used RL algorithms nor considered different traffic types.

In this chapter, we formulate the considered problem as a continuous-time multi-armed bandit (CT-MAB) model [6]. The unique characteristics of CT-MAB lie in that each time playing an arm takes a random period of time and the goal is to maximize the expected reward received in one unit time, which is suitable for balancing QoE reward and transmission cost in network selection. A traffic-aware online network selection algorithm (ONES) is designed based on an existing algorithm in [6], which, however, is limited by slow convergence speed due to multi-task learning. Therefore, we notice the correlation feature among multiple traffic types and utilize it to propose another two other efficient algorithms: D-ONES and VM-ONES. The main results of this chapter were presented in [7].

3.2   System Model

Similar to Chap. 2, it is assumed that a multimode user equipment (UE) locates in the overlapped coverage area of M NAP $\mathcal{M} = \{m_1, m_2, \ldots, m_M\}$ in HWN. The network selection scheme operates in a slot-based manner. Upon the arrival of a traffic, such as transferring a file, the UE will adjust the accessed network $a \in \mathcal{M}$ and start a transmission in the transmission procedure, based on a given network selection policy π. Note that the slot length here may last for several seconds and each traffic may last for multiple slots.

QoE is used as the network selection metric since it can accurately reflect the overall acceptability of an application or service. For the transmission in each slot, the UE evaluates the QoE reward ε and the transmission/network access cost τ, which are important feedback information for countering the NSI uncertainty and realizing online network selection algorithms in our scheme.

The QoE reward is derived by QoE functions. Explicitly, the QoE functions map the experienced NSI into the user's QoE reward in each slot. Considering different types of traffic or user demand in the UE, we define specific QoE function for each type of traffic according to its characteristic. While the traffic type classification can be diverse depending on actual considerations, there are some widely used classifications in the existing literature. Basically, according to the characteristic of throughput requirement, traffic is classified as stream traffic and elastic traffic in [8]. The stream traffic can only partially benefit from the allocated throughput, which includes audio and video applications. The elastic traffic can benefit from all the throughput such as web browsing, e-mail, file transfer. The authors in [1] further incorporate the brittle traffic into consideration and get three types of traffic: brittle traffic, stream traffic, and elastic traffic. The brittle traffic represents traffic that places a strict requirement on bandwidth and has no adaptive properties, which may include video telephony, telemedicine, etc.. On the other hand, according to the running applications, traffic is classified into video traffic, audio traffic, and file transfer in [9, 10]. We do not specify a concrete traffic classification, instead, we assume that the preferred traffic type set are known and denoted by $S = \{s_1, s_2, \ldots, s_{|S|}\}$, where $s \in S$ is one of $|S|$ traffic types. The stationary probability of each arriving traffic being type $s \in S$ is p_s with $0 < p_s < 1$ and $\sum_{s \in S} p_s = 1$. The QoE function for traffic type s $D\left(\mathcal{V}\right)$ maps experienced NSI into QoE reward, i.e., the QoE reward

$$\varepsilon = D_s\left(\mathcal{V},\right) \tag{3.1}$$

where $\mathcal{V} = \{v_1, v_2, \ldots, v_{|\mathcal{V}|}\}$ is the related NSI, $|\mathcal{V}|$ is the number of parameters. The user "experienced NSI" indicates that the NSI is the feedback from an end-to-end transmission perspective rather than the estimated or probed result. Actually, the effective sets of parameters for different traffic types may be subsets of $\mathcal{V}$ and can be different, as in [9].

3.3 Problem Formulation

We formulate the online network selection problem as a CT-MAB problem in this section.

3.3.1 QoE Reward Rate

Based on the observed history information $\Lambda\left(i\right) = \{s_1, a\left(1\right), \varepsilon\left(1\right), \tau\left(1\right), \ldots, s_{i-1}, a\left(i - 1\right), \varepsilon\left(i - 1\right), \tau\left(i - 1\right)\}$, a network selection policy π makes a decision, i.e., $a\left(i\right) = \pi\left(\Lambda\left(i\right)\right)$, where s_i, $a\left(i\right)$, $\varepsilon\left(i\right)$, and $\tau\left(i\right)$ are the traffic type, the selected network, the QoE reward, and network access cost of the ith slot, respectively. We assume that the NSI is approximately fixed during one slot. Actually, there is no need to store all the history information, since our algorithms work in an online updating manner.

Intuitively, the UE's goal is to find a network selection policy maximizing the expectation of the accumulative QoE reward, for instance, the expected total QoE rewards in successive T slots, $E\left[\sum_{i=1}^{T} D_{s_i}\left(\mathcal{V}\left(i\right)\right)\right]$, where $\mathcal{V}\left(i\right)$ is the experienced NSI in ith slot. However, this goal is not appropriate in our setting. Indeed, the accumulative QoE reward cannot capture the influence of network access cost. For example, a policy with the maximal accumulative QoE reward may incur unbearable access cost, such as energy consumption or transmission fee, and is thus not preferred. Alternatively, since the reward rate, which is defined as the ratio of reward and cost, can provide fair tradeoff between QoE reward and network access cost, we resort to maximize the expected QoE reward rate g_π:

$$
\begin{aligned}
g_\pi &= \lim_{T \to \infty} \sup E \left[\frac{\sum_{i=1}^{T} D_{s_i} (\mathcal{V}(i))}{\sum_{i=1}^{T} \tau(i)} \right] \\
&= \lim_{T \to \infty} \sup E \left[\frac{\sum_{i=1}^{T} \varepsilon(i)}{\sum_{i=1}^{T} \tau(i)}, \right]
\end{aligned}
\tag{3.2}
$$

where E means the expectation under policy π.

Considering $\varepsilon(i)$ and $\tau(i)$, the optimal network selection policy is determined by the characteristics of traffic as well as NSI. The traffic type s_i determines the distinct QoE function, the NSI $\mathcal{V}$ affects the QoE reward and the network access cost. Therefore, the traffic type together with the unknown and dynamic NSI make the computation of g_π challenging. It is possible that we can perform online learning on the QoE reward rate of each policy from communication and find the optimal policy. However, learning the optimal policy faces the exploring and exploiting dilemma. Maximizing user's QoE needs more access to networks estimated to be optimal, while converging to the optimal strategy requires sufficient access to networks estimated to be suboptimal. Therefore, tradeoff between exploitation and exploration of the heterogeneous wireless network resource is important.

3.3.2   Continuous-Time MAB Formulation

The above QoE reward rate maximization problem can be well modeled in the continuous-time MAB (CT-MAB) framework. We outline two key features differentiating the CT-MAB problem [6] from classical MAB as introduced in Chap. 2: *side information and cost*. Specifically, playing an arm also incurs cost, which is a random variable related to the side information and selected arm. The side information is a set of "task type" information indicating some particular utility for each paly. Therefore, the optimal arm(s) for different side information may be different. These differences result in the goal of the CT-MAB problem, maximizing the reward rate, while in the classical bandit problem the goal is maximizing the accumulative expected reward.

To model the problem, the available wireless networks are treated as the arms. Selecting a network corresponds to playing an arm, where the QoE reward ε and the network access cost τ in each slot are the reward and cost, respectively. Different traffic types have diverse QoE functions, thus the traffic type $s \in S$ of each slot is exactly the side information. In the rest of this chapter, we use the traffic type and side information interchangeably. If the rewards and costs in different slots are independent, maximizing the expectation in (3.2) is equivalent to maximizing the expectations' ratio of the numerator and the denominator [11], that is

$$g_\pi = \lim_{T \to \infty} \sup \frac{E_\pi \left[\sum_{i=1}^{T} \varepsilon(i) \right]}{E_\pi \left[\sum_{i=1}^{T} \tau(i) \right]}. \tag{3.3}$$

For the above CT-MAB problem, the optimal network selection policy π^* is the one maximizing the expected QoE reward rate

$$(\text{P1}) \quad \pi^* = \arg \sup_\pi g_\pi. \tag{3.4}$$

3.4 CT-MAB Based Network Selection Algorithms

In this section, the property of the optimal network selection policy is analyzed, based on which three RL- based network selection algorithms are then proposed.

3.4.1 Property of the Optimal Network Selection Policy

The goal of CT-MAB is to maximize the average reward rate, thus, classical UCB1 [12] and the similar learning algorithms cannot be directly used. However, learning algorithms for CT-MAB can be obtained by some transformations of the goal.

Denote $\varepsilon_{m,s}(i)$ and $\tau_{m,s}(i)$ as the reward and cost when the traffic type is s and network m is selected in the ith slot, respectively. We assume that the rewards and costs are bounded by

$$\varepsilon_{m,s}(i) \in [\varepsilon_{\min}, \varepsilon_{\max}], \tau_{m,s}(i) \in [\tau_{\min}, \tau_{\max}.]$$

The expected reward and cost when network m is selected with traffic type s are

$$E_{m,s} = E\left[\varepsilon_{m,s}(i)\right], \Gamma_{m,s} = E\left[\tau_{m,s}(i)\right]$$

. Denote the relative value $\kappa_{m,s}$ as the expected reward that can be collected minus the expected reward the optimal policy can collect when network m is selected with side information s

$$\kappa_{m,s} = E_{m,s} - \Gamma_{m,s} g^*, \tag{3.5}$$

where $g^* = \sup_{\pi} g^{\pi}$. Based on the theory of semi-Markov decision problems [6], the following result holds.

Theorem 1 *A deterministic stationary policy* $\pi^* : S \to M$ *is optimal for the CT-MAB problem, if and only if it satisfies the condition*

$$\kappa_{\pi^*(s),s} = \max_{m \in M} \kappa_{m,s} \quad \forall s \in S, \tag{3.6}$$

where $\pi(s)$ *denotes the selected network for traffic type s under policy* π.

Denote Π as the set of stationary deterministic policy, thus $|\Pi| = |M|^{|S|}$. This theorem indicates that the optimal network selection policy is $\pi^* \in \Pi$ that chooses a network maximizing the relative value $\kappa(m, s)$ for each traffic type $s \in S$ simultaneously. Therefore, (P1) can be changed to (P2):

$$(P2) \; \pi^* = \left\{ \pi(s) \mid \pi(s) = \arg\max_{m \in M} \kappa_{m,s}, \forall s \in S \right\}. \tag{3.7}$$

In the following, we first propose an online network selection algorithm, ONES. After analyzing its performance, we propose two more efficient algorithms, D-ONES and VM-ONES, by exploiting the correlation between learning tasks.

3.4.2 Online Network Selection Algorithm: ONES

Referring to the existing learning algorithm in [6], we are able to propose an online network selection algorithm (ONES) in Algorithm 3. The main idea of this algorithm is to develop the upper estimates of $\kappa_{m,s}$ with particular confidence bounds. The algorithm is as follows: in the first $|M|$ slots of each traffic type $s \in S$, each network $m \in M$ is selected once. After that, the network maximizing the upper estimate of κ_{m,s_i} is selected for each arriving traffic type s_i. The parameters in the algorithm are defined as

$$T_{m,s}(i) = \sum_{t=1}^{i} I(a(t) = m, s_t = s), \tag{3.8}$$

$$T_s(i) = \sum_{t=1}^{i} I(s_t = s), \tag{3.9}$$

$$T_{\pi}(i) = \sum_{t=1}^{i} I(a(t) = \pi(s_t)), \tag{3.10}$$

where $I(\cdot)$ is the indicator function, $T_{m,s}(i)$ is the number of slots when network m is selected for traffic type s in the first i slots. $T_s(i)$ is the number of slots when the traffic type is s in the first i slots. $T_\pi (i)$ is the number of slots when the selected network is compatible with policy π, that is, the same with the decision of policy π. In addition,

$$\bar{\varepsilon}_{m,s} (i) = \frac{1}{T_{m,s} (i)} \sum_{t=1}^{i} I (a (t) = m, s_t = s) \varepsilon (t) \tag{3.11}$$

$$\bar{\tau}_{m,s} (i) = \frac{1}{T_{m,s} (i)} \sum_{t=1}^{i} I (a (t) = m, s_t = s) \tau (t) \tag{3.12}$$

$$\bar{g}_\pi (i) = \frac{\sum_{t=1}^{i} I (a (t) = \pi (s_t)) \varepsilon (t)}{\sum_{t=1}^{i} I (a (t) = \pi (s_t)) \tau (t)} \tag{3.13}$$

$$\bar{g}^* (i) = \max_{\pi \in \Pi} \left[\bar{g}_\pi (i) - c_{i,T_\pi (i)}, \right] \tag{3.14}$$

where $\bar{\varepsilon}_{m,s} (i)$ and $\bar{\tau}_{m,s} (i)$ are the average reward and average cost in the first ith slots when network m is selected for traffic type s, respectively. $\bar{g}_\pi (i)$ is the estimate of average reward rate of network selection policy π in ith slot and $\bar{g}^* (i)$ is a lower estimate of g^* in ith slot, where

$$c_{i,T_\pi (i)} = \sqrt{2c_1 \log \left(i \sqrt{|\Pi| + 1} \right) / T_\pi (i)}, \tag{3.15}$$

$$c_1 = 2 \max \left\{ \frac{(\varepsilon_{\max} - \varepsilon_{\min})^2}{\tau_{\min}^2}, \frac{\varepsilon_{\max}^2 (\tau_{\max} - \tau_{\min})^2}{\tau_{\min}^4} \right\}.$$

Algorithm 3 ONES

1: **Initiate:** $i=0$, $T_s (0) = 0$, $\bar{\varepsilon}_{m,s} (0) = 0$, $\bar{\tau}_{m,s} (0) = 0$ for $m \in \mathcal{M}$, $s \in \mathcal{S}$.
2: **loop**
3:     Upon the arrival of ith traffic,
4:     **if** $T_{s_i} (i - 1) \leq |\mathcal{M}|$ **then**
5:         Select $\{T_{s_i} (i - 1) + 1\}$th network in $\mathcal{M}$.
6:     **else**
7:         Select network $a (i)$ to access and transmit.

$$a (i) = \arg \max_{m \in \mathcal{M}} \bar{\varepsilon}_{m,s_i} (i - 1) - \bar{\tau}_{m,s_i} (i - 1) \bar{g}^* (i - 1) + \hat{c}_{i-1,T_{m,s_i} (i-1)} \tag{3.16}$$

8:     **end if**
9:     Update $\bar{\varepsilon}_{m,s_i} (i)$, $\bar{\tau}_{m,s_i} (i)$, $\bar{g}^* (i)$ and $\hat{c}_{i,T_{m,s_i} (i)}$ according to (3.11)–(3.18).
10: **end loop**

In ONES,

$$\bar{\varepsilon}_{m,s_i}(i-1) - \bar{\tau}_{m,s_i}(i-1)\,\bar{g}^*(i-1) + \hat{c}_{i-1,T_{m,s_i}(i-1)} \tag{3.17}$$

is the upper estimate of κ_{m,s_i} and $\hat{c}_{i,T_{m,s}(i)}$ is the confidence interval defined as

$$\hat{c}_{i,T_{m,s}(i)} = (d_0 + d_1)\sqrt{\log\left(i\sqrt{|\Pi|+1}\right)/T_{m,s}(i)} \tag{3.18}$$

$$d_1 = \sqrt{2\tau_{\max}^2 c_1},$$

$$d_0 = \sqrt{8\max\left\{(\varepsilon_{\max} - \varepsilon_{\min})^2,\ \varepsilon_{\max}^2\,(\tau_{\max} - \tau_{\min})/\tau_{\min}^2\right\}}.$$

The regret is a well-known performance metric for online learning algorithms. Different from classical regret, the regret in our problem can be defined as

$$R(i) = g^* \sum_{t=1}^{i} \tau(i) - \sum_{t=1}^{i} \varepsilon(i) \tag{3.19}$$

Note that to maximize $\kappa_{m,s}$ in (P2) is equivalent to minimize the regret. Generally, the optimal grow speed of the regret is demonstrated to be the logarithmic order of the play number. The optimal logarithmic regret of ONES is guaranteed by the following theorem.

Theorem 2 (Theorem 1 in [6]) *The regret $R(i)$ of ONES in i slots is upper bounded as*

$$E[R(i)] \leqslant G^*\left[\left(2 + \frac{2\pi^2}{3\,(|\Pi|+1)^2}\right)|\mathcal{M}|\,|\mathcal{S}| + 2\,|\mathcal{M}|\,|\mathcal{S}|\log(i)\right.$$

$$\left. + \sum_{m:\Delta_m>0}\sum_{s\in S}\frac{(d_0+d_1)\log\left(i\left(\sqrt{|\Pi|+1}\right)\right)}{\Delta_m(s)^2}\right] \tag{3.20}$$

where $G^* = \tau_{\max}g^* - \varepsilon_{\min}$, $\Delta_m(s) = \kappa_{\pi^*(s),s} - \kappa_{m,s}$.

3.4.3 Decoupled Online Network Selection Algorithm: D-ONES

Besides the proposed ONES, other algorithms could also be derived when a different perspective is adopted. Formally, we can treat (P2) as a compound MAB problem, which consists of $|\mathcal{S}|$ sub-MAB problems. The sub-MAB problems are distinguished by the traffic type $s \in S$. The available networks $\mathcal{M}$ are the common $|\mathcal{M}|$ arms of all sub-MAB problems. Each sub-MAB problem with traffic type s aims at finding the optimal network to maximize $\kappa_{m,s}$. The aim of the compound MAB problem is to find the network selection policy maximizing $\kappa_{m,s}$ for $s \in S$, simultaneously. Consequently, the compound MAB requires multi-task online learning.

After analyzing the details of ONES, we can find that the update of sub-MABs in ONES are correlated. Note that i in $\hat{c}_{i,T_i(m,s)}$ is the total slot number of all the traffic rather than the slot number of traffic type s, $T_s(i)$. This results in a relative larger confidence interval $\hat{c}_{i,T_i(m,s)}$ than that in conventional MABs, e.g., in [12]. The relative large confidence intervals lead to conservative updating of sub-MABs, where the sampling in suboptimal networks for the online network selection is more emphasized. Hence, the correlated updating of sub-MABs directly induces performance loss.

In order to improve the efficiency of ONES, we try to seek for a new algorithm with a finer upper estimate for $\kappa_{m,s}$. Fortunately, Theorem 1 indicates that this multi-task online learning can be decomposed by the relative values, which provides the possibility. To this end, we propose a new algorithm, decoupled online network selection algorithm (D-ONES, see Algorithm 2), which decouples the updating of sub-MABs.

Denote

$$\phi_{t,t'} = \sqrt{\frac{2\log t}{t'}}, \tag{3.21}$$

we get an upper confidence estimate of $E_{m,s}$ as

$$\bar{\varepsilon}_{m,s}(i) + \phi_{T_s(i),T_{m,s}(i)},$$

and the lower confidence estimate of $\Gamma_{m,s}$ as

$$\bar{\tau}_{m,s}(i) - \tau_{\max}\phi_{T_s(i),T_{m,s}(i)}.$$

Since $\bar{g}_i^*$ is a lower estimate of g^*, we obtain a new upper estimate of $\kappa_{m,s}$ in ith slot as

$$\bar{\varepsilon}_{m,s}(i) + \phi_{T_s(i),T_{m,s}(i)} - \left[\bar{\tau}_{m,s}(i) - \tau_{\max}\phi_{T_s(i),T_{m,s}(i)}\right]\bar{g}_i^*. \tag{3.22}$$

Different from ONES, the upper estimate of $\kappa_{m,s}$ in D-ONES is related with $T_{i,s}$, rather than i. Note that given the same $T_{m,s}(i)$, the confidence interval in (3.22) is smaller than that in (3.17), which means that the decoupled updating results in a finer upper estimate in D-ONES. Thus, we can expect a better performance in D-ONES. Also, the following theorem demonstrates the logarithmic order regret of D-ONES under a mild condition.

Algorithm 4 VM-ONES

1: **Initiate:** $i=0$, $T_s(0) = 0$, $\bar{\varepsilon}_{m,s}(0) = 0$, $\bar{\tau}_{m,s}(0) = 0$ for $m \in \mathcal{M}$, $s \in \mathcal{S}$.
2: **loop**
3:     Upon the arrival of ith traffic,
4:     **if** $T_{s_i}(i-1) \leq |\mathcal{M}|$ **then**
5:         Select $\{T_{s_i}(i-1)+1\}$th network in $\mathcal{M}$.
6:     **else**
7:         Select network $a(i)$ to access and transmit.

$$a(i) = \arg\max_{m \in \mathcal{M}} \bar{\varepsilon}_{m,s_i}(i-1) + \phi_{T_{s_i}(i-1), T_{m,s_i}(i-1)} \tag{3.23}$$

$$- \left[\bar{\tau}_{m,s_i}(i-1) - \tau_{\max}\phi_{T_{s_i}(i-1), T_{m,s_i}(i-1)}\right]\bar{g}^*(i-1)$$

8:     **end if**
9:     Update $\bar{\varepsilon}_{m,s_i}(i)$, $\bar{\tau}_{m,s_i}(i)$ and $\bar{g}^*(i)$ according to (3.11)–(3.14), update $d_{T_{s_i}(i-1), T_{m,s_i}(i-1)}$ according to (3.21).
10: **end loop**

Theorem 3 *The regret $R(i)$ of D-ONES in i slots is upper bounded as*

$$E[R(i)] \leq \sum_{s \in \mathcal{S}} p_s L^*(s) \sum_{m \neq \pi^*(s)} E\left[\sum_{t=1}^{i} I\{\pi_t(s) = m\}\right]$$

$$= \sum_{s \in \mathcal{S}} p_s L^*(s) \sum_{m \neq \pi^*(s)} \left[8\zeta_{m,s}\log t + 1 + \frac{\pi^2}{3}\right] \tag{3.24}$$

$$\zeta_{m,s} = \max_{\bar{g}_{i-1}^*} \frac{\left(1 + \bar{g}_{i-1}^* \tau_{\max}\right)^2}{\left[E_{\pi^*(s),s} - E_{m,s} + \bar{g}_{i-1}^*\left(\Gamma_{m,s} - \Gamma_{\pi^*(s),s}\right)\right]^2},$$

when the following (3.25) holds for $\forall m \neq \pi^*(s)$, $s \in \mathcal{S}$

$$E_{\pi^*(s),s} - E_{m,s} + \bar{g}_{i-1}^*\left(\Gamma_{m,s} - \Gamma_{\pi^*(s),s}\right) \geq 0, \tag{3.25}$$

where p_s is the stationary probability of each arriving traffic being type $s \in \mathcal{S}$, $L^*(s) = \max_{m \in \mathcal{M}} \{\Gamma_{m,s}g^* - E_{m,s}\}$ is the largest loss for traffic type s.

The proof of Theorem 3 can be seen in the appendix. Note that

$$\Delta_m(s) = \kappa_{\pi^*(s),s} - \kappa_{m,s} = E_{\pi^*(s),s} - E_{m,s} + g^*\left(\Gamma_{m,s} - \Gamma_{\pi^*(s),s}\right) > 0. \tag{3.26}$$

When i is sufficiently large, $\bar{g}_{i-1}^* \to g^*$ with high probability, which means that (3.25) holds. Therefore, the above regret bound can be achieved with high probability.

3.4.4 *Virtual Multiplexing Online Network Selection Algorithm: VM-ONES*

Comparing ONES and D-ONES, we can find that they share some commons on the algorithm update manner. As we have mentioned above, one sub-MAB problem with side information the same as current traffic type is updated in each slot for both ONES and D-ONES. Accordingly, the bounds of the regrets of the two algorithms are approximately the sum average regret of each sub-MAB. This can be seen in the bound of $E\left[R\left(i\right)\right]$ in (3.20) where $\left(2 + \frac{2\pi^2}{3(|\Pi|+1)}\right)|\mathcal{M}|\,|\mathcal{S}|$ and $2\,|\mathcal{M}|\,|\mathcal{S}|\log\left(i\right)$ grow linearly with $|\mathcal{S}|$, i.e., the number of sub-MABs. Similar result can be found in the regret bound of D-ONES. Although the regrets of ONES and D-ONES grow in the optimal logarithmic order, their actual regrets are still considerably large.

With this observation, it is possible to further promote the learning efficiency. Along a different avenue with ONES and D-ONES, we consider a new approach to exploit the correlation among sub-MABs to speed up the updating, rather than decoupling the updating of sub-MABs. Actually, there exist available correlation among sub-MABs. We notice that for a selected network $a\left(i\right) = m'$ in ith slot with traffic type $s_i = s'$, the average QoE reward $\bar{\varepsilon}_{m',s}\left(i\right)$ and average network access cost $\bar{\tau}_{m',s}\left(i\right)$ can be updated not only for sub-MAB with side information $s = s'$ but also for sub-MABs with $s \neq s'$. This is possible if for fixed network m' and experienced NSI $\mathcal{V}\left(i\right)$, the only difference among $\varepsilon_s\left(i\right)$ and $\tau_s\left(i\right)$ for $s \in \mathcal{S}$ lies in the different QoE functions of the traffic. This motivates us to utilize the correlation in such cases with an appropriate formulation.

Algorithm 5 VM-ONES

1: **Initiate:** $i{=}0$, $T_s\left(0\right) = 0$, $\bar{\varepsilon}_{m,s}\left(0\right) = 0$, $\bar{\tau}_{m,s}\left(0\right) = 0$ for $m \in \mathcal{M}, s \in \mathcal{S}$.
2: **loop**
3:    Upon the arrival of ith traffic,
4:    **if** $T_{s_i}\left(i-1\right) \leq |\mathcal{M}|$ **then**
5:       Select $\left\{T_{s_i}\left(i-1\right)+1\right\}$th network in $\mathcal{M}$.
6:    **else**
7:       Select network $a\left(i\right)$ to access and transmit.

$$a\left(i\right) = \arg\max_{m \in \text{mathcal}M}\,\bar{\varepsilon}_{m,s_i}\left(i-1\right) - \bar{\tau}_{m,s_i}\left(i-1\right)\bar{g}^*\left(i-1\right) + \hat{c}_{i-1,T_{m,s_i}\left(i-1\right)} \tag{3.27}$$

8:    **end if**
9:    **loop**
10:       Create virtual samples $\{s, \tilde{a}_s\left(i\right), \tilde{\varepsilon}_s\left(i\right), \tilde{\tau}_s\left(i\right)\}$ for $s \in \tilde{\mathcal{S}}_i$.
11:    **end loop**
12:    For each $s \in \mathcal{S}$, update $\bar{\varepsilon}_{m,s}\left(i\right)$ and $\bar{\tau}_{m,s}\left(i\right)$ according to (3.29)–(3.30), update $\bar{g}^*\left(i\right)$ and $\hat{c}_{i,T_{m,s}\left(i\right)}$ according to (3.14) and (3.18) with virtual samples.
13: **end loop**

We update all the sub-MABs in each slot by constructing virtual samples $\{s, \tilde{a}_s\left(i\right), \tilde{\varepsilon}_s\left(i\right), \tilde{\tau}_s\left(i\right)\}$ for $s \neq s_i$. Denoting the virtual side information set in ith slot as $\tilde{S}_i = \{s \mid s \neq s_i, s \in S\}$, we assume that there are virtual traffic (type $s \in \tilde{S}_i$) arrivals with the current traffic (type s_i) at the same time and they can virtually multiplex in the selected network, which is $\tilde{a}_s(i) = m'$. The virtual network access cost is $\tilde{\tau}_s(i) = \tau(i)$ and the QoE reward can be updated as

$$\tilde{\varepsilon}_s(i) = D_s(\mathcal{V}(i)), \forall s \in \tilde{S}_i. \tag{3.28}$$

The sample means of the rewards and the access costs are updated as follows:

$$\bar{\varepsilon}_{m,s}(i) = \frac{\sum_{t=1}^{i} I(a(t)=m, s_t=s)\varepsilon(t) + \sum_{t=1}^{i} I\left(\tilde{a}_s(t)=m, s \in \tilde{S}(t)\right)\tilde{\varepsilon}_s(t)}{T_{m,s}(i) + \tilde{T}_{m,s}(i).} \tag{3.29}$$

$$\bar{\tau}_{m,s}(i) = \frac{\sum_{t=1}^{i} I(a(t)=m, s_t=s)\tau(t) + \sum_{t=1}^{i} I\left(\tilde{a}_s(t)=m, s \in \tilde{S}(t)\right)\tilde{\tau}_s(t)}{T_{m,s}(i) + \tilde{T}_{m,s}(i)} \tag{3.30}$$

where $\tilde{T}_{m,s}(i) = \sum_{t=1}^{i} I\left(\tilde{a}_s(i) = m, s \in \tilde{S}_i\right)$ is the number of virtual samples with traffic type s and selected network m observed up to the ith slot. The sample means of the QoE reward and the access cost incorporate the virtual samples and all the sub-MABs can be updated in parallel. Similarly, the virtual samples should also be incorporated in computing $\bar{g}_\pi(i)$. Based on the virtual samples, we propose a virtual multiplexing online network selection algorithm (VM-ONES, see Algorithm 3). Interestingly, we found that the regret of VM-ONES has the same bound as ONES. The proof is similar with the proof of Theorem 1. Nevertheless, the simulation and following analysis indicate a significant reduction in the actual regret of VM-ONES, compared with ONES.

3.4.5 Convergence Performance Analysis

We approximately compare the convergence performance of VM-ONES with that of ONES in a special case, where each traffic lasts for only one slot. With virtual samples, there are one real traffic (type s_i) and $|S| - 1$ virtual traffic types, which means that all the $|S|$ sub-MABs can be updated in parallel in each slot. Suppose that there is only one type of traffic s in the network selection problem, and in each slot, the sub-MAB with s is updated in ONES. Denote $W(p^*)$ as the state that the probability of selecting the best network is no lower than p^*. We use n_s, the average slot number for single sub-MAB to converge to state $W(p^*)$, to reflect the convergence speed. For the compound MAB of the network selection problem, ONES and VM-ONES converge to $W(p^*)$ only when all the sub-MABs converge to $W(p^*)$. In ONES, the probability that sub-MAB with s is updated is p_s and the total slot number for sub-MAB with s to converge is approximately $\frac{n_s}{p_s}$. Therefore, the average slot number for ONES to converge to $W(p^*)$ is

$$N_{ONES} = \max_{s \in S} \frac{n_s}{p_s}.$$

On the other hand, all $|S|$ sub-MABs can be updated in parallel in VM-ONES. Denote $n_{s_0} = \max\limits_{s \in S} n_s$, the average slot number for VM-ONES to converge to $W(p^*)$ is

$$N_{VM\text{-}ONES} = \max_{s \in S} n_s = n_{s_0}.$$

Since $\max\limits_{s \in S} \frac{n_s}{p_s} \geq \frac{n_{s_0}}{p_{s_0}}$, thus, the convergence performance gain

$$\frac{N_{ONES}}{N_{VM\text{-}ONES}} \geq \frac{1}{p_{s_0}}.$$

Accordingly, the convergence performance advantage of VM-ONES can effectively decrease the regret.

3.5 Simulation Results

In this section, some simulation results are presented to validate the proposed algorithms.

3.5.1 Simulation Setting

Similar to Chap. 2, it is assumed that a multimode UE can access three NAPs: WLAN1, WLAN2, and LTE. Due to the complexity in NSI dynamics, we adopt a discrete model similar to [5] to model the packet loss rate, delay, and throughput of networks. The packet loss probability of the network $e_{network}$ in a slot can be any of the following N_e states:

$$e_{n_e} = e_{\min} + e_{unit} n_e, \quad n_e = 1, \ldots, N_e$$

where $e_{\min}$ is the minimal packet loss rate, e_{unit} is a packet loss rate unit, and N_b is the number of state. For example, when $e_{\min} = 0.001$, $e_{unit} = 0.0005$ and $N_b = 3$, then the packet loss can be approximated into three states as 0.0015, 0.002, and 0.0025. Similarly, the network delay $d_{network}$ is characterized by $d_{\min}$, d_{unit}, and N_d. The throughput θ is characterized by $\theta_{\min}$, θ_{unit}, and N_θ. Accordingly, the instant NSI can be represented by the joint state (e, d, θ).

Three types of traffic similar with [9] are considered: s_v-video traffic, s_a-audio traffic, and s_e-elastic traffic. The corresponding traffic type set is denoted as $S = \{s_v, s_a, s_e\}$. We emphasize that more complicated user demands could also be adopted.

Some widely used QoE models are specified for each type of traffic. The QoE reward in our simulation is the mean opinion score (MOS) [13], which is used as a subjective measure of the network quality. The MOS has five values from 1 to 5 indicating users' satisfactory degrees "Bad", "Poor", "Fair", "Good", and "Excellent", respectively. In the following, we briefly present the QoE functions of the three traffic types, which map user experienced NSI into MOS output.

For video traffic, the MOS is mainly dependent on the loss of a single slice of a frame from the video stream [9]. With some transformation, the MOS is simplified as a function of the Peak Signal-to-Noise Ratio (PSNR), the QoE function D_{video} is

$$D_{video}\left(P_{snr}\right) = 4.5 - \frac{3.5}{1 + \exp\left(b_1\left(P_{snr} - b_2\right)\right)}, \tag{3.31}$$

where b_1 and b_2 are the parameters determining the shape of the function, P_{snr} is the experienced PSNR. In the simulation, $b_1 = 1$ and $b_2 = 5$.

For audio traffic, the QoE function D_{audio} is defined by a nonlinear mapping of the R-factor [14]

$$D_{audio}\left(R_f\right) = 1 + 0.035 R_f + 7 \cdot 10^{-6} R_f\left(R_f - 60\right)\left(100 - R_f\right), \tag{3.32}$$

where R_f is the R-factor defined by ITU to reflect the audio quality impairment from different aspects. Generally, delay and packet loss rate are two main concerns. Thus, R_f can be computed by [14]

$$R_f = 94.2 - I_e - I_d$$

where I_e is the impairment caused by packet loss rate, I_d is the impairment caused by delay. Given a packet loss rate e, the impairment I_e is defined as [14, 15]

$$I_e = \gamma_1 + \gamma_2 \ln\left(1 + \gamma_3 e\right)$$

where γ_1, γ_2 and γ_3 are constant parameters dependent on the codec. It is recommended that when G.729a is used, the parameters are $\gamma_1 = 11$, $\gamma_2 = 40$ and $\gamma_3 = 10$. The packet loss rate e consists of the loss probability in the network $e_{network}$ and the loss probability caused by playout loss $e_{playout}$ as $e = e_{network} + \left(1 - e_{network}\right)e_{playout}$. In the simulation, $e_{playout} = 0.005$. On the other hand, the delay impairment I_d reflects the effect caused by the delay d in ms,

$$I_d = 0.024 d + 0.11\left(d - 177.3\right) I\left(d - 177.3\right)$$

where $I\left(\cdot\right)$ is the indicator function. Note that $177.3\,\text{ms}$ is believed to be a delay threshold for audio traffic [15]. The delay d consists of codec delay d_{codec}, playout delay $d_{playout}$ and network delay $d_{network}$ as $d = d_{codec} + d_{playout} + d_{network}$, where $d_{codec} = 25\,\text{ms}$, $d_{playout} = 60\,\text{ms}$ are adopted in the simulation.

Table 3.1 Parameters in the simulation

	e_{min}	e_{unit}	N_e	d_{min} (ms)	d_{unit} (ms)	N_d	θ_{min} (kbps)	θ_{unit} (kbps)	N_θ	P_{snr} (dB)
LTE	0.002	0.002	3	20	10	7	250	50	11	4
WLAN1	0.002	0.002	4	50	10	6	400	50	18	5
WLAN2	0.002	0.002	5	60	10	5	250	50	11	7

For non-real-time traffic such as file transfer and web browsing, we call it as elastic traffic. The corresponding QoE is defined as an increasing function of throughput θ [16]

$$D_{elastic}(\theta) = b_3 \log(b_4\theta,) \tag{3.33}$$

where b_3 and b_4 can be determined by the required maximal and minimal throughput. We assume the required minimal and maximal throughput are 100 and 2000 kbps and the resulting parameters are set as $b_3 = 2.6949$, $b_4 = 0.0235$.

The parameters in Table 3.1 are used mainly referring to both the reference in [17] and our trace data by android application "speedtest". The PSNR is assumed fixed for each network. The network access cost τ here refers to nominal transmission fee. We assume that the WLAN is closed access manner and its transmission fee is slightly more than that of LTE. Specifically, the costs are set as 1.1, 1.2, and 1.2 for LTE, WLAN1, and WLAN2, respectively. The transmission power or energy could also be set as the cost. For the traffic type in each slot, the stationary probabilities of being video traffic, audio traffic, and elastic traffic are $P_1 = 0.5$, $P_2 = 0.3$, and $P_3 = 0.2$, respectively.

3.5.2 Convergence Behavior

We first give the sample runs of ONES, D-ONES, and VM-ONES. In order to clearly see the network selection trend, each traffic only lasts for one slot. In Figs. 3.1, 3.2, and 3.3, the x-axis of each figure represents the slot index and the y-axis represents the ratio of selecting a network for some specific traffic type. We can observe in Fig. 3.1 that, as the slot number increases, the achieved deterministic network selection policy in ONES is π_{ONES} : {video $\rightarrow$ WLAN2, audio $\rightarrow$ LTE, elastic $\rightarrow$ LTE} . Figures 3.2 and 3.3 illustrate that D-ONES and VM-ONES converge to the same network selection policy as ONES, i.e., $\pi_{D-ONES} = \pi_{VM-ONES} = \pi_{ONES}$. To check whether the achieved deterministic stationary network selection policy is optimal, we simulated 10000 slots averaged by 100 runs for each of the $\mathcal{M}^{|S|} = 27$ deterministic stationary network selection policies to derive the average QoE reward rate. The result shows that the above network selection policy indeed achieves the largest QoE reward rate, i.e., 3.27.

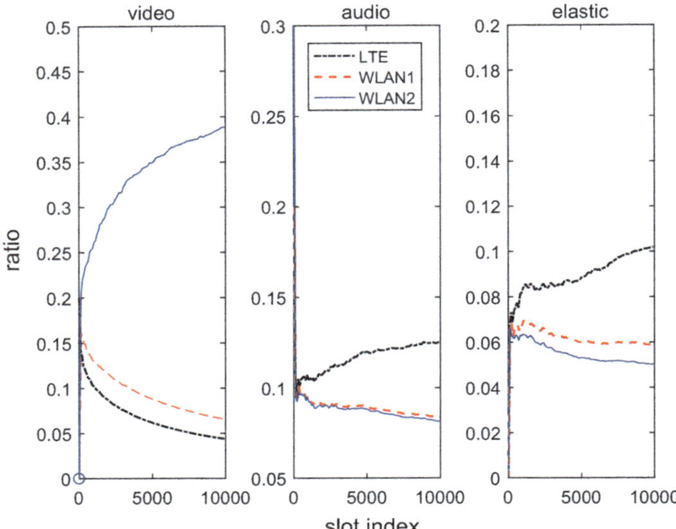

Fig. 3.1 Sample run of ONES

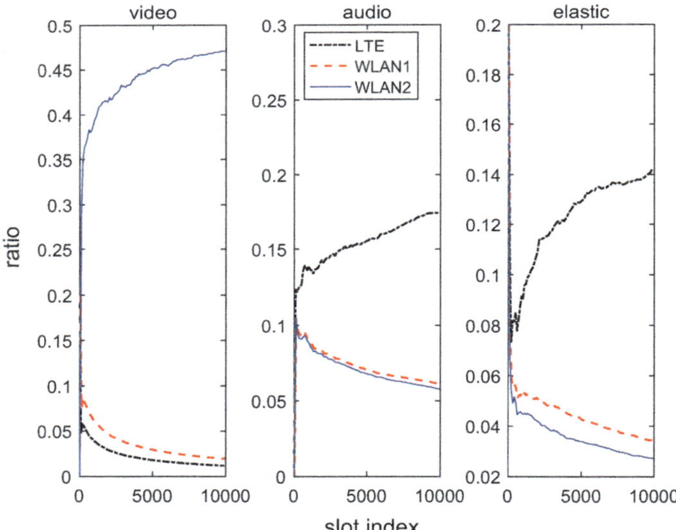

Fig. 3.2 Sample run of D-ONES

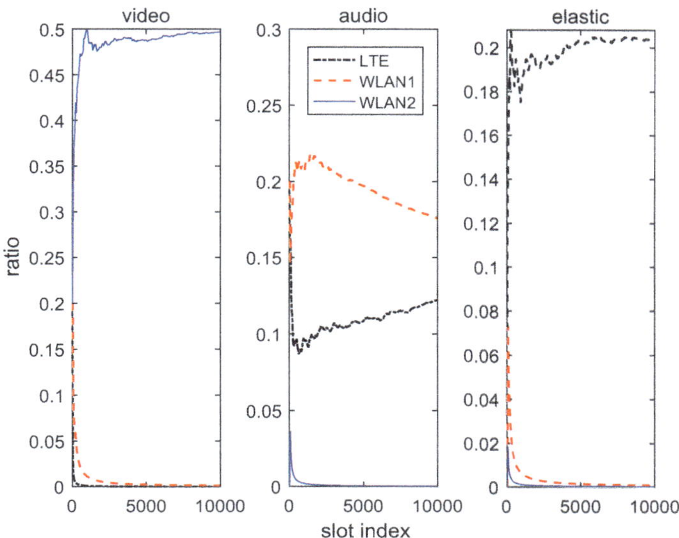

Fig. 3.3 Sample run of VM-ONES

We point out that even though it takes about 100 slots for D-ONES and VM-ONES to achieve an outstanding reward rate, this does not hamper their applications. On one hand, the proposed algorithms are online selections rather than selection without communication. The key insight of this work is to leverage the data transmission process for learning, where performance improvements can be achieved throughout the process. On the other hand, the algorithms do not incur any additional cost except for the lightweight computation cost. It is worth noting that although the ratio of selecting LTE is not the largest for audio traffic in VM-ONES in Fig. 3.3, its trend indicates LTE is optimal. The reason behind this phenomenon may be: since RL algorithm has to explore WLAN1 with sufficient samples, given that it is not selected frequently for video and elastic traffic, it has to balance the sample number, i.e., sampling LTE more frequently.

3.5.3 Performance Comparison

We compare the regrets of the three algorithms in Fig. 3.4 in two scenarios. In the following simulation, each traffic lasts for a random slot number in the range of [1, 10]. The regrets are obtained by averaging the regrets of 100 runs and each run is simulated in 10000 slots. We can see that the regrets of three algorithms grow in a sub-linear order of slot number. In addition, both D-ONES and VM-ONES have

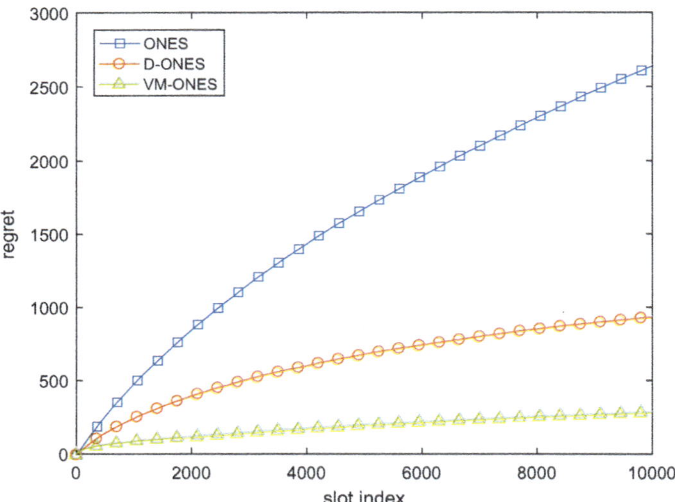

Fig. 3.4 Regrets of the three algorithms

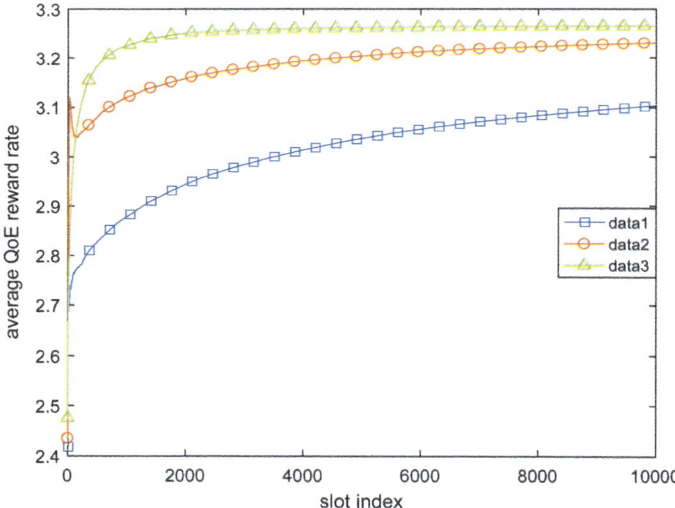

Fig. 3.5 Average QoE reward rates of the three algorithms

much smaller regrets than ONES, and VM-ONES achieves the smallest regret. This verifies the performance improvement of D-ONES and VM-ONES.

Accordingly, Fig. 3.5 gives the average QoE reward rates of three algorithms. As we can expect, VM-ONES achieves the largest QoE reward rate and ONES achieves the smallest. We can also find that the performance gap is mainly due to convergence

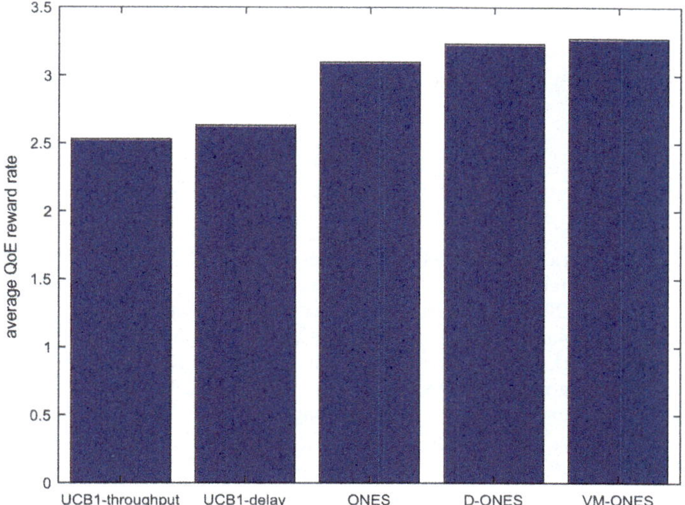

Fig. 3.6 Performance comparison with learning algorithms without QoE awareness

speed diversity, because the three algorithms tends to converge to some QoE reward rate bound. The results validate the importance of convergence speed.

Finally, two learning-based network selection algorithms without QoE consideration and traffic awareness are considered. The first one seeks for the network offering the largest average throughput, where UCB1 algorithm is adopted to learn the throughput optimal network. We denote this algorithm as UCB1-throughput. The second algorithm adopts UCB1 to learn the network with the smallest average delay, denoted as UCB1-delay. We omit the details of these two algorithms, since they can be easily constructed with some modifications to the original UCB1 algorithm in [12]. The average QoE reward rates at the 10000th slot averaged by 100 runs of all algorithms are shown in Fig. 3.6. As can be seen that the QoE reward rates of UCB1-delay and UCB1-throughput are significantly smaller than those of the proposed three algorithms due to the lack of QoE consideration and traffic awareness.

3.6 Conclusions

This chapter studied the cost–performance ratio optimization with dynamic traffic types in dynamic and uncertain HWN. Considering the transmission cost, we formulated the problem as a CT-MAB model to balance the QoE reward rate and transmission cost. Three CT-MAB RL algorithms were proposed. To improve the learning efficiency issue in multi-task learning due to various traffic types, two algorithms, D-ONES and VM-ONES were designed based on the existing ONES algorithm. The regret bound of D-ONES is theoretically analyzed and simulation results demonstrated the effectiveness as well as the performance improvements of QoE awareness of the three algorithms.

References

1. Trestian R, Ormond O, Muntean G (2012) Game theory-based network selection: solutions and challenges. IEEE Commun Surv Tut 2(99):1–20
2. Hou J, Brien DCO (2006) Vertical handover-decision-making algorithm using fuzzy logic for the integrated Radio-and-OW system. IEEE Trans Wirel Commun 5(1):176–185
3. Martinez-Morales JD, Pineda-Ricoand U, Stevens-Navarro E (2010) Performance comparison between MADM algorithms for vertical handoff in 4G networks. In: Proceedings of the 7th international conference on electrical engineering computing science and automatic control (CCE)
4. Piamrat K, Ksentini A, Viho C et al (2008) QoE-based network selection for multimedia users in IEEE 802.11 wireless networks. In: IEEE local computer networks (LCN)
5. Stevens-Navarro E, Lin Y, Wong VWS (2008) An MDP-based vertical handoff decision algorithm for heterogeneous wireless networks. IEEE Trans Veh Technol 57(2):1243–1254
6. György A, Kocsis L et al (2007) Continuous time associative bandit problems. In: Proceedings of the 20th international joint conference on artificial intelligence (IJCAI)
7. Wu Q, Du Z, Yang P, Yao Y, Wang J (2016) Traffic-aware online network selection in heterogeneous wireless networks. IEEE Trans Veh Technol 65(1):381–397
8. Haddad M, Elayoubi SE, Altman E et al (2011) A hybrid approach for radio resource management in heterogeneous cognitive networks. IEEE J Sel Areas Commun 29(4):831–842
9. Reis AB, Chakareski J, Kassler A et al (2010) Distortion optimized multi-service scheduling for next-generation wireless mesh networks. In: IEEE INFOCOM
10. Manzoor A, Umar T (2011) User utility function as quality of experience (QoE). In: The 10th international conference on networks (ICN)
11. Ferguson TS (2008) Optimal stopping and applications. http://www.math.ucla.edu/~tom/Stopping
12. Auer P, Cesa-Bianchi N, Fischer P (2002) Finite-time analysis of the multiarmed bandit problem. Mach Learn 47:235–256
13. Anandkumar A, Michael N et al (2011) Distributed algorithms for learning and cognitive medium access with logarithmic regret. IEEE J Sel Areas Commun 29(4):731–745
14. ITU-T Recommendation G.107 (1998) The E-model: a computational model for use in transmission planning. https://www.itu.int/rec/T-REC-G.107
15. Sengupta S, Chatterjee M, Ganguly S (2008) Improving quality of VoIP streams over WiMax. IEEE Trans Mobile Comput 57(2):145–156
16. Kelly FP (1997) Charging and rate control for elastic traffic. Eur Trans Telecommun 8:33–37
17. Raychaudhuri D, Mandayam NB (2012) Frontiers of wireless and mobile communications. Proc IEEE 100(4):824–840

Part II
MDP RL Based Online Network Selection

This part studies online network selection from the perspective of Markov decision process (MDP). Different from MAB in the previous part, MDP is able to provide an analysis framework for more fine-grained optimization, i.e., *state*-dependent decision-making. This motivates us to introduce MDP for realizing user demand and context-aware online network selection. In Chap. 4, dynamic user demand and network handoff are jointly formulated as an MDP and efficient RL algorithms are proposed. This model is further extended to the dynamic context case in Chap. 5 and transfer RL algorithm is introduced to promote the learning efficiency by reusing experience. Both chapters formulate network selection as MDP, the algorithm design focusses on how to improve the learning efficiency, especially for context with large space.

Chapter 4
Meeting Dynamic User Demand with Handoff Cost Awareness: MDP RL Based Network Handoff

Abstract This chapter focuses on the network selection for dynamic user demand. The evolution of traffic types indicates dynamic user demands, which may result in user demand specific optimal networks. Considering the network handoff cost, learning the optimal matching between user demand and networks has to tradeoff QoE reward and handoff cost in the long-term learning process. We formulate this problem as a Markov decision process (MDP), where network handoff cost is quantitatively reflected as loss in QoE reward. By exploiting the prior information on the problem, we propose two efficient Q-learning-based algorithms. Simulation results indicate that the proposed algorithms can well solve the online matching between dynamic user demand and network with handoff cost awareness.

4.1 Introduction

This chapter studies the impact of network handoff on learning considering dynamic user demand. Nowadays, smartphones support different traffic types and can run various applications which impose different performance requirements on transmission. In this context, user demand would become dynamic as the traffic/application type evolves. For instance, users may switch among watching video clips, making an audio call, and browsing webpages. Intuitively, the best strategy is to learn the optimal matching between network and user demand. However, the changing of user demand may lead to frequent network handoffs. In extreme cases, the handoff cost even may erode the performance gain achieved from the network and user demand matching. Therefore, there is a tradeoff between the network selection and network handoff.

In the literature, the traffic types and user demand have been considered in some related works [1, 2], while little focus is on their dynamic feature. Although network handoff cost is considered in many vertical handoff-related works, it is not studied from the learning perspective. For example, the authors in [3, 4] try to maximize the expected total reward of a connection subject to the expected total access cost constraint. The admission of users is formulated as a semi-Markov decision process

© Springer Nature Singapore Pte Ltd. 2020
Z. Du et al., *Towards User-Centric Intelligent Network Selection
in 5G Heterogeneous Wireless Networks*,
https://doi.org/10.1007/978-981-15-1120-2_4

and a value iteration algorithm is used to obtain an optimal stationary deterministic policy in [5]. These approaches are based on dynamic programming but not learning algorithms. The network selection is formulated as learning problems but the user demand is not considered in [6]. A Q-learning-based algorithm is proposed to select the best network based not only on the current network state but also the potential future network and device states in [7]. A distributed user association algorithm for network load balancing in vehicular networks is used in [8]. Chapter 2 of this book has studied the network handoff in learning, but it did not consider dynamic user demand. In addition, network handoff is not quantitatively modeled but is just a concern for learning algorithm design.

In this chapter, the network handoff/selection for dynamic user demand is formally modeled as a Markov decision process (MDP). Compared to MAB in Chap. 2, MDP enables more fine-grained learning and optimization, suitable for characterizing user demand dynamics. For algorithm part, a Q-learning-based algorithm is proposed. Moreover, two efficient Q-learning-based algorithms that exploit some prior information to improve standard Q-learning are proposed. The main results of this chapter were presented in [9].

4.2 System Model and Problem Formulation

We assume that a user locates in the overlapping area of N wireless networks $\mathcal{N} = \{1, 2, \ldots, N\}$. In a slotted system with slot duration l second, the user can dynamically change its access network at the beginning of each slot. Similar to the previous two chapters, the NSI of all networks is dynamic and unknown to the user. In addition, since the user may pose different performance requirements and even the dominating QoS parameters in user perception may vary with time, the dynamics of user demand is considered.

Denote the user demand type set as $\mathcal{S} = \{s_1, \ldots, s_L\}$. We assume that the user demand is invariant during one slot and may change between two successive slots. Specifically, the dynamic of user demand between two successive slots follows some unknown Markov transition matrix $\mathbf{P}(l) = \{p_{s,s'}\}$, $s, s' \in \mathcal{S}$, where $p_{s,s'}$ is the one step transition probability when the user demand changes from s to s' given the slot duration l. Different slot durations l may result in different transition matrices. For each type of user demand $s \in \mathcal{S}$, there is a distinct demand utility function $D_s(\mathbf{v})$ mapping the experienced QoS to users' utility, e.g., satisfaction degree or QoE, where $\mathbf{v}$ is the relevant QoS parameter vector. The network handoff cost matrix is $\mathbf{C} = \{c_{m,n}\}$, $m, n \in \mathcal{N}$, where $c_{m,n}$ is the cost for the handoff from network m to n. Here, $c_{m,n}$ even can be random variables with bounded mean and variance.

Denote the currently associated network at the beginning of tth slot by $n(t)$ and the network selection decision for tth slot by $\delta(t)$. Jointly considering the user's QoE and network handoff cost, the user's reward $u(t)$ is

$$u(t) = D_{s(t)}(\mathbf{v}(t)) - \lambda c_{\delta(t),n(t)}, \tag{4.1}$$

where $\lambda \in [0, 1]$ is the handoff cost weight determining the tradeoff between the QoE and handoff cost, $\mathbf{v}(t)$ and $s(t)$ are the experienced QoS parameter vector in network $\delta(t)$ and the user demand type in tth slot, respectively. We noticed that the relationship of the current associated network and the network selection decision is $n(t+1) = \delta(t)$. Due to the time-varying user demand $s(t)$ and dynamic $\mathbf{v}(t)$, it is reasonable to optimize the cumulative discounted expected reward R. Formally, given an initial user demand type $s(0)$ and network $n(0)$, we have to find a sequence of network selection decisions $\{\delta(0), \delta(1), \ldots\}$ to maximize

$$R(s(0), n(0)) = E\left[\sum_{t=0}^{\infty} \beta^t u(t) \,|\, s(0), n(0)\right] \tag{4.2}$$

where the discount factor $\beta \in (0, 1)$ controls the future rewards' effect on the cumulative reward.

The concern of network handoff makes the network selection learning even more challenging than that in Chap. 1. Apparently, selecting the best network for each specific traffic type or user demand is optimal. However, such a requirement may incur frequent network handoffs due to the dynamic user demand, which calls for elaborate design. To this end, we try to solve the network selection problem in a Markov decision process (MDP) framework [10]. Generally, an MDP can be characterized by state, action, reward, and state transition matrix. We map the problem into an MDP as follows:

- State: $\mathbf{x}(t) = [s(t), n(t)]$. The state space is $\mathbf{X} = \mathcal{S} \times \mathcal{N}$.
- Action: $\delta(t) \in \mathcal{N}$.
- Reward: $r(t) = u(t) = u(\mathbf{x}(t), \delta(t))$.
- State transition matrix: $\mathbf{P}' = \{p(\mathbf{x}'|\mathbf{x}, \delta)\}, \mathbf{x}, \mathbf{x}' \in \mathbf{X}, \quad p(\mathbf{x}'|\mathbf{x}, \delta) = \begin{cases} p_{s_{\mathbf{x}'}, s_{\mathbf{x}}}, & n_{\mathbf{x}'} = \delta \\ 0, & otherwise \end{cases}$,
  where $s_{\mathbf{x}}$ and $n_{\mathbf{x}}$ are the corresponding user demand type and associated network in state $\mathbf{x}$, respectively.

The system state $\mathbf{x}(t)$ here is jointly determined by the user demand type and the associated network. As a result, the action can affect the state transition. It is worth noting that the action space is part of the state space, which is different from most existing MDP models. Given the finite states and actions, we define a stationary and deterministic policy $\pi \in \Pi$ as a function mapping the states to actions $\mathbf{X} \to \mathcal{N}$, where Π is the space of stationary and deterministic policy. Then, the problem in (4.2) can be changed to find a policy satisfying

$$\pi^* = \arg\max_{\pi \in \Pi} E_\pi\left[\sum_{t=0}^{\infty} \beta^t r(t) \,|\, \mathbf{x}(0)\right] \tag{4.3}$$

where E_π means taking expectation under policy π.

4.3 Algorithm Design

In practical implementations, the prior user demand transition probability matrix $\mathbf{P}(l)$ and the NSI dynamics are unknown, making the direct optimization of (4.3) such as dynamic programming, intractable. To get around this limitation, we resort to reinforcement learning. In the following, two online network selection algorithms based on Q-learning are proposed.

4.3.1 Parallel Q-Learning Algorithm 1

The first algorithm is based on the classical Q-learning algorithm. Due to limited space, we omit the details of Q-learning and related information can be seen in [10]. The main idea of Q-learning is to learn the Q value $Q(\mathbf{x}, \delta)$ for each state–action pair, which is the estimated expected long-term reward for action δ in state $\mathbf{x}$. The action for each state is selected according the updated Q values. Although standard Q-learning based online network selection algorithm can converge to the optimal policy, it may be complex and slow to convergence. In order to achieve fair performance, each state–action pair $Q(\mathbf{x}, \delta)$ must be sampled sufficiently. In our problem, there are $|\mathcal{S}|\,|\mathcal{N}|$ states and $|\mathcal{N}|$ actions, resulting in $|\mathcal{S}|\,|\mathcal{N}|^2$ state–action pairs. The relatively large state–action pair space may lead to poor convergence performance of Algorithm 1. The slow convergence speed may be one limitation in practical implementations.

Fortunately, we found that there exists room to speed up Q-learning algorithm. The dependence among utilities in different states enables us to update a group of state–action pairs simultaneously. If certain prior information about the system is known, we can speed up the leaning process.

Condition 1: *The network handoff cost* $c_{m,n}$, $m, n \in \mathcal{N}$ *is constant and known.*

We can notice that if condition 1 is satisfied, for some state $\mathbf{x} = [s, n]$ and the action δ, we can update $|\mathcal{N}|$ state–action pairs $r_i = u(\mathbf{x}_i, \delta)$, $\forall \mathbf{x}_i \in \mathbf{X}(s_\mathbf{x})$, where $\mathbf{X}(s_\mathbf{x}) = \left\{ \mathbf{x}' \,\middle|\, \forall \mathbf{x}' \in \mathbf{X}, s_{\mathbf{x}'} = s_\mathbf{x} \right\}$ is the set of states whose user demand type is the same with $\mathbf{x}$. Note that the differences of $u(\mathbf{x}_i, \delta)$ for $\forall \mathbf{x}_i \in \mathbf{X}(s_\mathbf{x})$ only lie in the handoff cost according to (4.1). Thus, it makes sense to update the state–action pairs within group $\mathbf{X}(s_\mathbf{x})$. Therefor, we propose a parallel Q-learning Algorithm 1. It works in a similar process as standard Q-learning: At initiation, $Q(\mathbf{x}, \delta)$ is set with zero for each state–action pair. Then, for each state, the network δ maximizing the corresponding Q value is selected. Finally, the immediate reward r and the new state are used to update the Q value, where α_t is the learning rate satisfying $\sum_{t=0}^{\infty} \alpha_t = \infty$, $\sum_{t=0}^{\infty} \alpha_t^2 < \infty$ [10]. A key difference is that we defined a new matrix G as the same size with Q to facilitate parallel update in (4.4) and (4.4)

Algorithm 6 Parallel Q Learning Algorithm 1

1: **Initiate:** $t = 0$, $Q(\mathbf{x}, \delta) = 0$ and $G(\mathbf{x}, \delta) = 0 \ \forall \mathbf{x} \in \mathbf{X}, \forall \delta \in \mathcal{N}$.
2: **loop**
3:   At the beginning of t-th slot, observe state $\mathbf{x}(t)$ and select network $\delta(t)$ according to the following rule:
4:   • With probability ε, randomly select $\delta(t) \in \mathcal{N}$;
5:   • Else, $\delta(t) \in \arg\max_{\delta \in \mathcal{N}} Q(\mathbf{x}(t), \delta)$.
6:   Receive reward $r(t)$ at the end of slot t.
7:   The system state changes to $\mathbf{x}(t+1) = [s(t+1), \delta(t)]$
8:     Update as follows

$$G(\mathbf{x}, \delta(t)) = (1 - \alpha_t) Q(\mathbf{x}, \delta(t)) + \alpha_t \left[r(t) + \beta \max_{\delta \in \mathcal{N}} Q(\mathbf{x}(t+1), \delta) \right], \forall \mathbf{x} \in \mathbf{X}\left(s_{\mathbf{x}(t)}\right)$$

$$(4.4)$$

$$Q(\mathbf{x}, \delta(t)) = G(\mathbf{x}, \delta(t)), \forall \mathbf{x} \in \mathbf{X}\left(s_{\mathbf{x}(t)}\right) \tag{4.5}$$

9: **end loop**

4.3.2 Parallel Q Learning Algorithm 2

We also found that some other prior information could further speed up the learning.

- **Condition 2**: *The network handoff cost* $c_{m,n}$, $m, n \in \mathcal{N}$ *is constant and known.*
- **Condition 3**: *QoE functions are known and all the QoS parameters in* $\mathbf{v}$ *are observable in each slot.*

When Conditions 2 and 3 are satisfied, we can obtain $|\mathcal{S}||\mathcal{N}|$ state–action pairs $r_i = u(\mathbf{x}_i, \delta)$, $\forall \mathbf{x}_i \in \mathbf{X}$. Note that the above samples comprise one real sample $\mathbf{x}_i = \mathbf{x}$ and $|\mathcal{S}||\mathcal{N}| - 1$ virtual samples. The virtual samples are derived based on the fact that the difference among $u(\mathbf{x}_i, \delta)$ for $\forall \mathbf{x}_i \in \mathbf{X}$ lies in the demand utility $D_s(\mathbf{v})$ and the handoff cost. If all the QoS parameters in $\mathbf{v}$ are observable in each epoch, the group update manner is feasible. Therefore, we propose parallel Q-learning Algorithm 2, whose only difference to Algorithm 1 is replacing (4.4) and (4.4) with (4.6) and (4.6), respectively.

4.4 Simulation Results

In this section, some simulation results and trace-based results are presented.

4.4.1 Simulation Setting

We assume that three networks are available for the user: LTE, WLAN1, and WLAN2. Although the instant SNR changes with time, the dynamics of peak SNRs can be slow

---

Algorithm 7 Parallel Q Learning Algorithm 2

1: **Initiate:** $t = 0$, $Q(\mathbf{x}, \delta) = 0$ and $G(\mathbf{x}, \delta) = 0$ $\forall \mathbf{x} \in \mathbf{X}$, $\forall \delta \in \mathcal{N}$.

2: **loop**

3:   At the beginning of t-th slot, observe state $\mathbf{x}(t)$ and select network $\delta(t)$ according to the following rule:

4:   • With probability ε, randomly select $\delta(t) \in \mathcal{N}$;

5:   • Else, $\delta(t) \in \arg\max_{\delta \in \mathcal{N}} Q(\mathbf{x}(t), \delta)$.

6:   Receive reward $r(t)$ at the end of slot t.

7:   The system state changes to $\mathbf{x}(t+1) = [s(t+1), \delta(t)]$

8:     Update as follows

$$G(\mathbf{x}, \delta(t)) = (1 - \alpha_t) Q(\mathbf{x}, \delta(t)) + \alpha_t \left[r(t) + \beta \max_{\delta \in \mathcal{N}} Q(\mathbf{x}(t+1), \delta) \right], \forall \mathbf{x} \in \mathbf{X} \quad (4.6)$$

$$Q(\mathbf{x}, \delta(t)) = G(\mathbf{x}, \delta(t)), \forall \mathbf{x} \in \mathbf{X} \quad (4.7)$$

9: **end loop**

Table 4.1 Parameter set

	e_m	e_u	K_e	d_m	d_u	K_d	θ_m	θ_u	K_θ
LTE	0.02	0.02	3	10	10	5	250	50	6
WLAN1	0.02	0.02	5	50	10	4	450	60	4
WLAN2	0.04	0.02	5	60	10	5	250	50	4

and bounded. Thus, we assume that peak SNRs of networks are fixed throughout for simplicity. In realistic wireless communications, the dynamics of QoS parameters such as the delay, loss, and throughput can be very complex. According to the trace data from Android application "speedtest" [11], we found that the Markov chain model in [3] can approximately model the dynamics of QoS parameters. Specifically, the joint delay–loss–throughput state is fixed in one slot, while state transition at successive two slots follows fixed probability. The parameters are set as shown in Table 4.1 and the transition probabilities between any two states are assumed equal. In the table, the delay d in ms, loss e, and throughput θ in kbps are all characterized by three parameters: subscript "m" denotes the minimal values, subscript "u" denotes the basic unit, K denotes the number of levels. The nominal handoff cost matrix among LTE, WLAN1, and WLAN2, the user demand transition matrix among $video, audio$, and $elastic$ are

$$\mathbf{C} = \begin{matrix} 0 & 2 & 3 \\ 2 & 0 & 1 \\ 3 & 1 & 0 \end{matrix}, \mathbf{P}_0 = \begin{matrix} 0.4 & 0.3 & 0.3 \\ 0.3 & 0.2 & 0.5 \\ 0.5 & 0.2 & 0.3 \end{matrix}$$

The video traffic demand, audio traffic demand, and elastic traffic demand denoted by $\mathcal{S} = \{video, audio, elastic\}$ are considered. The QoE functions are the same with those in Chap. 2. In the simulation, $\lambda = 0.5$, $\beta = 0.3$, and $\lambda = 0.6$.

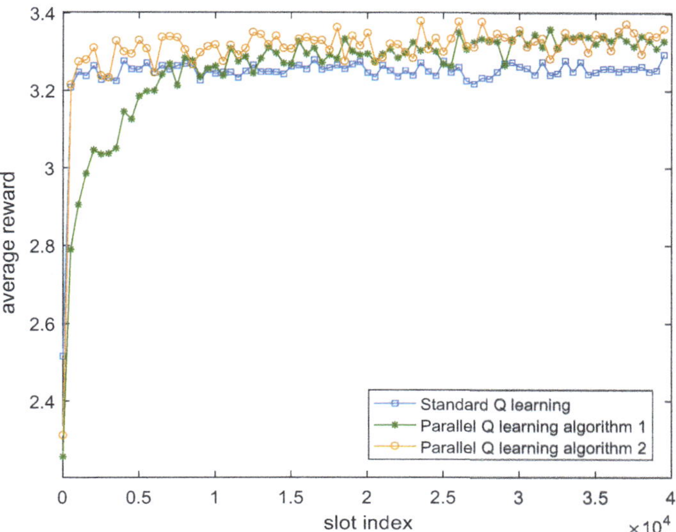

Fig. 4.1 Convergence behavior of different algorithms

4.4.2   *Results*

The convergence performance of the proposed two algorithms and standard Q-learning based algorithm are presented in Fig. 4.1. As can be seen, the proposed two algorithms can achieve much larger average rewards than the standard Q-learning based algorithm. This indicates that the improved update manner could effectively promote the convergence speed. Meanwhile, Algorithms 2 outperforms Algorithm 1 due to its large number of samples in the group update manner.

We compare the performance of the proposed algorithms with another five algorithms. The considered five algorithms include a random selection algorithm, a matching selection algorithm, and three fixed selection algorithms. The random selection algorithm randomly selects a network in each slot. The matching selection algorithm selects the optimal network according to the user demand. In our setting, matching selection means the optimal networks for video traffic demand, audio traffic demand, and elastic traffic demand are WLAN2, LTE, and WLAN1, respectively. The fixed selection algorithm indicates sticking to any one of networks without handoff. In Fig. 4.2, the average rewards in 1000 runs of different algorithms are shown. As can be seen, the performance of the random selection, the matching, and the proposed algorithm (proposed Algorithm 2) generally decrease as the handoff cost weight increases. In particular, since the random selection algorithm considers neither the user demand nor the handoff cost, its performance decays quickly and becomes the worst of all algorithms when $\lambda > 0.3$. On the other hand, since the matching algorithm selects the optimal networks for each type of user demand,

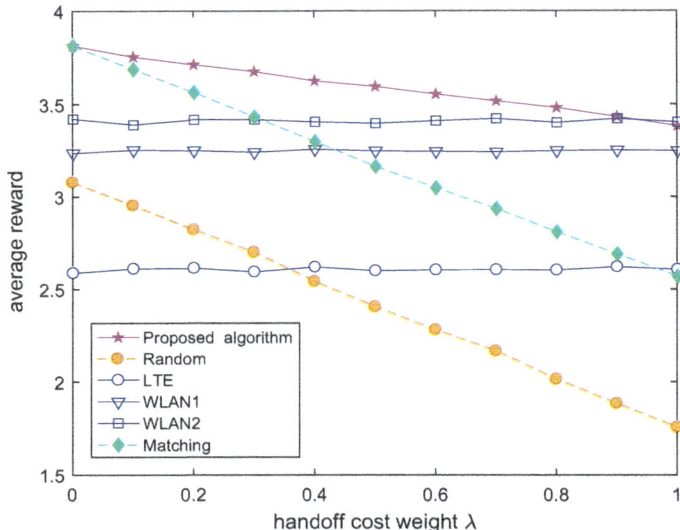

Fig. 4.2 Algorithm performance comparison with different handoff cost weight $\lambda \in [0, 1]$

its performance is the same with the proposed algorithms when $\lambda = 0$. However, when λ increases, its performance becomes worse than that of the proposed algorithm, mainly due to the negative influence of handoff cost. Even though the handoff cost decays the performance gain, the proposed algorithms outperform the other five algorithms.

The impact of user demand dynamics on the performance of the mentioned algorithms are further compared. Three different user demand transition matrices are considered: where $\mathbf{P}_1$ represents that the user demand prefers to maintain the current type with higher probability, $\mathbf{P}_2$ represents that the user demand can be any one of three types with equal probability in each slot, and $\mathbf{P}_3$ represents that the user demand changes with higher probability. In Fig. 4.3, we can find that although the performance of the proposed algorithms may vary in different cases, they are much better than the other algorithms.

$$\mathbf{P}_1 = \begin{matrix} 0.6 & 0.2 & 0.2 \\ 0.2 & 0.6 & 0.2 \\ 0.2 & 0.2 & 0.6 \end{matrix}, \mathbf{P}_2 = \begin{matrix} 0.34 & 0.33 & 0.33 \\ 0.33 & 0.34 & 0.33 \\ 0.33 & 0.33 & 0.34 \end{matrix}, \mathbf{P}_3 = \begin{matrix} 0.2 & 0.4 & 0.4 \\ 0.4 & 0.2 & 0.4 \\ 0.4 & 0.4 & 0.2 \end{matrix}$$

4.5 Conclusion

This chapter studied the network selection for dynamic user demand. Due to the diverse user demand requirements, we need to learn the optimal matching between user demand and networks, which, however, has to tradeoff QoE reward and handoff

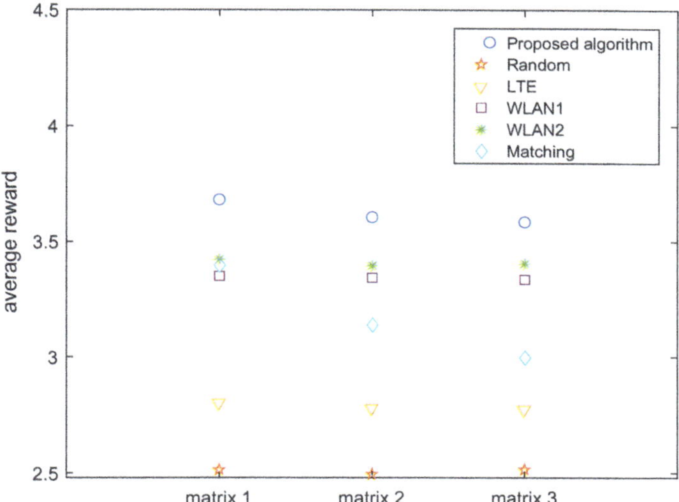

Fig. 4.3 Algorithm performance comparison with different transition matrices of user demand dynamics

cost in the long-term learning process. This tradeoff is solved in a MDP framework. Besides the standard Q learning, two parallel Q-learning algorithm based algorithms were proposed. Simulation results validated that the proposed algorithms can achieve better performance with much faster convergence speed.

References

1. Reis AB, Chakareski J, Kassler A et al (2010) Distortion optimized multi-service scheduling for next-generation wireless mesh networks. In: IEEE INFOCOM
2. Shamik S, Mainak C, Samrat G (2008) Improving quality of VoIP streams over WiMax. IEEE Trans Comput 57(8):145–156
3. Stevens-Navarro E, Lin Y, Wong VWS (2008) An MDP-based vertical handoff decision algorithm for heterogeneous wireless networks. IEEE Trans Veh Technol 57(2):1243–1254
4. Sun C, Stevens-Navarro E, Wong VWS (2008) A constrained MDP-based vertical handoff decision algorithm for 4G wireless networks. In: IEEE international conference on communications (ICC)
5. Kim TO, Devanarayana CN et al (2015) An optimal admission control protocol for heterogeneous multicast streaming services. IEEE Trans Commun 63(6):2346–2359
6. Haleh T, Golnaz F, John C (2011) A learning-based network selection method in heterogeneous wireless systems. In: IEEE global telecommunications conference (GLOBECOM)
7. Tabrizi H, Farhadi G, Cioffi J (2012) Dynamic handoff decision in heterogeneous wireless systems: Q-learning approach. In: IEEE international conference on communications (ICC)
8. Li Z, Wang C, Jiang C (2017) User association for load balancing in vehicular networks: an online reinforcement learning approach. IEEE Trans Intell Transp Syst 18(8):2217–2228

9. Du Z, Wu Q, Yang P (2014) Dynamic user demand driven online network selection. IEEE Commun Lett 18(3):419–422
10. Sutton RS, Barto AG (2017) Reinforcement Learning: An Introduction, 2nd edn. MIT Press, London
11. Speedtest. http://www.speedtest.net/

Chapter 5
Learning the Optimal Network with Context Awareness: Transfer RL Based Network Selection

Abstract This chapter focuses on how to realize context-aware network selection. The context information provides a characterization of user-, traffic-, and network-related properties, which is able to enable fine-grained optimization for network selection. Following a similar way with Chap. 4, we can formulate context-aware network selection as an MDP model by generalizing the state to context information. However, the high resolution of context information may lead to large state space, which could result in low learning efficiency. To handle this issue, we employ a transfer learning idea. Specifically, the time–location- dependent periodic changing rule of load statistical distributions is used to realize efficient online network selection via knowledge transfer. Simulation results show that the proposed transfer RL algorithm could achieve better convergence performance by reusing learning experience.

5.1 Introduction

The feature and dynamics of user demand is introduced in network selection in the previous chapter. Actually, the communication context of the user is dynamic, which inspires us to further take context information into account. While there is no formal definition of context, we consider three aspects that are important for network selection: user demand/traffic type, network performance, location, and time. In other words, we want to design fine-grained online network selection approach that is aware of the communication context of the user, aiming at ensuring high QoE.

In the literature, context-aware communication optimization has received much attention [1]. While traffic and application layer information is widely used to evaluate and compare network performance in network selection, however, the concept of context awareness is rarely mentioned. The authors in [2] considered the application types and hardware conditions as context information for small cell association and the problem is formulated as a matching game between small cell base stations and users. The optimal network selection policy for multiuser-integrated visible light

© Springer Nature Singapore Pte Ltd. 2020

Z. Du et al., *Towards User-Centric Intelligent Network Selection in 5G Heterogeneous Wireless Networks*,

https://doi.org/10.1007/978-981-15-1120-2_5

communication and positioning systems under unknown environment's dynamics is proposed in [3]. Although context information is considered in [4], to our knowledge, there is no related work on formally modeling context-aware network selection, especially for time- and location-information dependent network selection.

In this chapter, the communication context taking the location and time into account is considered in network selection. To realize context-aware network selection, the problem is formulated as an MDP problem. However, the relatively high resolution of context information may lead to a large state space, resulting in low learning efficiency. We exploit the fact that the mobility pattern of mobile users and the spatial–temporal distribution show strong regularity to reuse learning experience. To this end, we introduce the idea of transfer learning and propose reinforcement learning algorithms with knowledge transfer to promote the learning performance. The main results of this chapter were presented in [5].

5.2 System Model and Problem Formulation

We consider the HWN environment that consists of N networks $\mathcal{N} = \{1, 2, \ldots, N\}$ of LTE, WLAN, and visible light communication (VLC). For simplicity, NAP is to used to represent a base station (BS) in the LTE or an access point (AP) in the WLAN and VLC. In a slotted system with the epoch duration of l s, the user can dynamically change its access network, but only one access network can be accessed at any given time slot.

5.2.1 *Heterogeneous Network Performance*

We use throughput as the main performance metric of the networks. Many other performance metrics could be involved, which is beyond the focus of this chapter. The maximal instantaneous rate of a user that is determined by the SNR (signal-to-noise ratio) according to the Shannon formula constitutes the upper bound of its throughput. Meanwhile, the multiuser access behavior determines the real-time network load distribution and thus affects the achieved throughput of each user in the network. Therefore, the achieved throughput $\Theta\,(i, n)$ of user i in network n is a function of the instantaneous rate R and the network load K_n (the total number of users in network n) $\Theta\,(i, n) = f\,(R, K_n)$ for a given slot. The function $f\,(\cdot)$ could be modeled depending on the specific network. In the following, the uplink and downlink throughput models of LTE, WLAN, and VLC are given.

5.2.1.1 LTE

The OFDMA is the downlink multiple access technology of LTE. According to the model in [6], the throughput under weighted-proportional fairness can be expressed as

$$\Theta_{\mathrm{DL}}(i, n) = \frac{\omega_i R_{n \to i}}{W_k}, \tag{5.1}$$

where ω_i is user i's weight, $W_k = \sum_{i \in \mathcal{K}_n} \omega_i$ is the sum of weights of users, $\mathcal{K}_n$ is the set of users in network n and $K_n = |\mathcal{K}_n|$, $R_{n \to i}$ is the instantaneous downlink rate of user i.

In the uplink, LTE uses the SC-FDMA-based MAC protocol with fair subcarrier sharing. Hence, the throughput of user i is roughly dependent on the total number of users sharing the same network

$$\Theta_{\mathrm{UL}}(i, n) = \frac{R_{n \leftarrow i}}{K_n}, \tag{5.2}$$

where $R_{n \leftarrow i}$ is the instantaneous uplink rate of user i.

5.2.1.2 WLAN

In 802.11 WLAN MAC protocols, the distributed coordination function (DCF) leads to a fair access opportunity to uplink users. Hence, the low rate user capturing the channel will use it for a long time, thus penalizing high rate users. The uplink throughput of a WiFi user can be expressed as

$$\Theta_{\mathrm{UL}}(i, n) = \frac{L}{\sum_{j \in \mathcal{K}_n} \frac{L}{R_{n \leftarrow i}}}. \tag{5.3}$$

Here, L is the packet size. The throughput that a user can obtain on the downlink is related to the scheduling mechanism of the access point. According to [7], when a round-robin(RR) scheme is used, then the downlink throughput can also be derived by replacing $R_{n \leftarrow i}$ with $R_{n \to i}$ in formula (5.3).

5.2.1.3 VLC

We consider an all-optical VLC network. Downstream data transmission and illumination are combined. Currently, there is no common view on the MAC protocol specified for VLC. In most existing works, it is assumed that the system uses TDMA with RR scheduling. Thus, if user i is assigned to the nth VLC AP, the achieved throughput becomes [8]

$$\Theta_{\mathrm{DL}}(i, n) = \frac{R_{n \to i}}{2 \cdot K_n}. \tag{5.4}$$

Note that the intensity modulation with direct detection (IM/DD) is used in VLC and only real-valued signals can be transmitted to receivers. Thus, at least half of the subcarriers must be used to realize the Hermitian conjugate of the complex-valued symbol after modulation. Consequently, the formula is divided by 2.

Using visible light in uplink may not be practical since it would constrain equipment power and users' psychological feelings. Referring to [9], we use infrared in the uplink. The main limitation of the infrared link is its low power transmission, which often leads to a low data transmission rate (up to 4 Mbps or 1.152 Mbps in [10]). Since visible light and infrared light exhibit very similar qualitative behavior, the uplink throughput model could also be derived by replacing $R_{n \to i}$ with $R_{n \gets i}$ in formula (5.4).

5.2.2 *Heterogeneous Utility Function*

Considering the diverse features of various traffic types, we propose a general utility model to differentiate uplink and downlink performance requirements. Note that we mainly focus on the throughput, but this model can be easily extended to incorporate many other performance metrics. The achieved utility $u\left(\Theta_{\mathrm{UL}}, \Theta_{\mathrm{DL}}\right)$ is designed from a novel perspective.

5.2.2.1 Uplink-Dominated Traffic

For traffic such as sending files or backing up files on the cloud, the uplink throughput is the main factor affecting the performance. The downlink throughput is negligible since it is just for transmitting some control and feedback messages (no less than a small threshold, e.g., Θ_0). As an example, it can be defined using a similar utility representing file transfer.

$$u\left(\Theta_{\mathrm{UL}}, \Theta_{\mathrm{DL}}\right) = \mathrm{I}\left\{\Theta_{\mathrm{DL}} \geq \Theta_0\right\} \lambda log\left(\beta \cdot \Theta_{\mathrm{UL}}\right), \tag{5.5}$$

where $\mathrm{I}\{x\} = 1$ when $x = 1$, and otherwise $\mathrm{I}\{x\} = 0$. $\mathrm{I}\left\{\Theta_{\mathrm{DL}} \geq \Theta_0\right\}$ is the minimal downlink throughput requirement and $\lambda log\left(\beta \cdot \Theta_{\mathrm{UL}}\right)$ models the utility-throughput function [6].

5.2.2.2 Downlink-Dominated Traffic

On the contrary, downloading files and watching online videos mainly utilize the downlink throughput and can be classified as downlink dominant traffic. Since most existing works focus on this traffic type, the utility $u(\Theta_{\mathrm{DL}})$ can be easily derived by explicitly indicating the downlink throughput Θ_{DL} in existing utility models. For instance, the file download utility can use the above model by replacing Θ_{UL} with

Θ_{DL}. Video traffic shows a threshold effect on throughput. Then, a piecewise function of the downlink throughput plus the basic uplink throughput requirement is

$$u\left(\Theta_{\mathrm{UL}}, \Theta_{\mathrm{DL}}\right) = \begin{cases} 0 & \Theta_{\mathrm{DL}} \leq \Theta_1 \\ \frac{c(\Theta_{\mathrm{DL}} - \Theta_1)}{\Theta_2 - \Theta_1} \mathrm{I}\left\{\Theta_{\mathrm{UL}} \geq \Theta_0\right\} & \Theta_1 < \Theta_{\mathrm{DL}} < \Theta_2 \\ c\mathrm{I}\left\{\Theta_{\mathrm{UL}} \geq \Theta_0\right\} & \Theta_{\mathrm{DL}} \geq \Theta_2 \end{cases} \tag{5.6}$$

where c is a constant, and Θ_1 and Θ_2 are two throughput thresholds determined by the traffic requirements.

5.2.2.3 Uplink–Downlink Symmetric Traffic

Video calls and video conference traffic have high requirements on both the downlink and uplink throughput. Either uplink or downlink throughput can be the bottleneck. We can replace Θ_{DL} with $\Theta_{\min} = \min\left(\Theta_{\mathrm{UL}}, \Theta_{\mathrm{DL}}\right)$ in formula (5.6) to get a utility function.

5.2.3 *Reinforcement Learning Problem Formulation*

Due to channel fading and the shadowing effect, the instantaneous rates $R_{n \leftarrow i}(t)$ and $R_{n \rightarrow i}(t)$ are time varying. Moreover, the network load K_n is a random variable since the active user number in a network is dynamic. Consequently, the achieved throughput $\Theta(i, n)$ and the resulting $u\left(\Theta_{\mathrm{UL}}, \Theta_{\mathrm{DL}}\right)$ are dynamic and random variables. Hence, it is reasonable to select the network that provides the best average performance. However, since the user has no prior knowledge of the average performance of the available networks, he has to learn the optimal selection from the interaction with the environment. Mathematically, this learning problem can be formed to select a network selection policy π^* that maximizes the long-term average reward. In other words, it selects a series of actions $\{a(1), a(2), \ldots\}$ that can maximize the total expected return as

$$V^* = \max E\left\{\sum_{t=0}^{\infty} \gamma^t u\left[\Theta_{\mathrm{UL}}(a(t)), \Theta_{\mathrm{DL}}(a(t))\right]\right\}, \tag{5.7}$$

where $\gamma \in (0, 1)$ represents the discount factor that reflects future returns relative to their current importance. $u\left[\Theta_{\mathrm{UL}}(a(t)), \Theta_{\mathrm{DL}}(a(t))\right]$ is the instant reward received at time t, and $\Theta_{\mathrm{UL}}(a(t))$ and $\Theta_{\mathrm{UL}}(a(t))$ are the instant uplink and downlink throughput, respectively.

5.3 Algorithm Design

In this section, reinforcement learning based algorithm is proposed. In particular, context is formally defined and the idea of transfer learning is introduced to improve the learning efficiency.

5.3.1 Communication Context

The challenge of the above learning problem is the underlying communication context. Due to the mobility dynamics of mobile user, the communication context will evolve and the optimal network also varies, accordingly. This means that we have to learn context-dependent optimal network. Basically, this problem can be regarded as a large-scale constrained dynamic optimization problem embedded in a stochastic environment, where the communication context is the sate.

Similar to Chap. 4, reinforcement learning is one of the effective ways to find a solution. Differently, as communication context can include more information than just user demand, enabling more fine-grained network selection. Specifically, we introduce a vector $(s, \mathcal{N}^*, i)$ to represent the user demand(traffic type)–location–time contextual information, where $s \in \mathcal{S}$, $\mathcal{N}^* \subseteq \mathcal{N}$, and $i \in \mathcal{I}$ are the current user demand, available network set, and time period index, respectively. $\mathcal{S}$ is the set of traffic types, e.g., the three types defined in previous sections and $\mathcal{N}$ is the maximal available network set. Note that since the available networks may vary across different locations, we use the available network set to indicate the "location" instead of exact coordinates. This type of location label is an efficient location discrimination method tailored for network selection. One day is dividing into several time periods. For example, the daytime of weekdays from 8:00 am to 5:00 pm could be divided into 9 periods each corresponding to 1 h duration. The load statistical distributions of all networks are assumed to remain unchanged in each time period.

5.3.2 Reinforcement Learning with Knowledge Transfer

However, the standard Q-learning algorithm may show slow convergence speed and poor performance due to the exploration. When the available strategy set is relatively large, there will be significant random exploration costs on bad strategies. Especially, as the communication context is three dimensional, the induced state space is large. In the following, we attempt to reduce the learning complexity by exploiting the underlying features of the problem.

After analyzing the problem, we find the Q-learning can be enhanced with the following four observations.

- **Observation 1: Not all networks are inherently suitable for all traffic types**. There may be mismatches between the asymmetric downlink/uplink network performance and traffic requirements. For instance, VLC itself has poor uplink throughput due to the inherent limitations that we have mentioned. Thus, it is not suitable for the traffic with strict requirements on uplink performance.
- **Observation 2: Not all networks are preferred by the user for all scenarios**. The user may prefer certain networks. For instance, if the user frequently changes his/her posture or moves around the room, the smartphone could not maintain stable VLC access, and thus VLC may not be preferred. For fee consideration, the user may not want the LTE access due to its relatively high fees.
- **Observation 3: The mobility pattern of mobile users often show a high degree of temporal and spatial regularity**. It was revealed that mobility users can be characterized by a time-independent characteristic travel distance and a significant probability to return to a few highly frequented locations[11]. That is to say that mobile users generally visit among limited locations in daily life.
- **Observation 4: Network load statistical distribution is time–location dependent**. Recent literature [12] has revealed that the traffic/load shows the spatial and temporary distribution law. In other words, there is a periodic changing rule with the time of the load statistical distribution for a given location. This periodic changing rule regarding the load dynamics of networks may be used. For example, the load statistical distributions at a specific location at the same duration on different weekdays are generally the same.

Observations 1 and 2 enable us to decrease the size of action set according to the traffic type and user preferences. That is, some network choices can be removed in the Q-learning action set if they are not suitable. This is realized by selecting the traffic type-dependent action set, using the refined action set $\mathcal{N}_s \subseteq \mathcal{N}^*$ and the Q vector $Q = [Q\,(1)\,, Q\,(2)\,, \ldots, Q\,(|\mathcal{N}_s|)]$ as shown in the eighth line of the algorithm.

On the other hand, observation 3 and observation 4 actually indicate that the load statistical distribution at the same time period and the same location across different weekdays for mobile users are approximately the same. The idea of transfer learning [13] provides a feasible way to enhance the Q-learning algorithm via observation 3 and observation 4. The transfer learning transfers knowledge learned in certain *source tasks* and uses it to improve the efficiency of machine learning in a related *target task* apart from existing data/samples, as illustrated in Fig. 5.1. For reinforcement learning, the transfer learning enables us to accelerate the algorithm's convergence by using some knowledge or contextual information. One straightforward and effective transfer method is to set the initial solution in the target task based on a source task. In this way, the starting point of the learning process could be much closer to the final target task solution, compared to the standard reinforcement learning starting at fully randomly searching.

The next concern is how to define the source task and target task and map the two tasks. It is easy to find that the past learning experience that shares the same context $(s, \mathcal{N}^*, i)$ with the current learning task is the source task and the current learning task is the target task. Hence, the starting point of the current learning process could

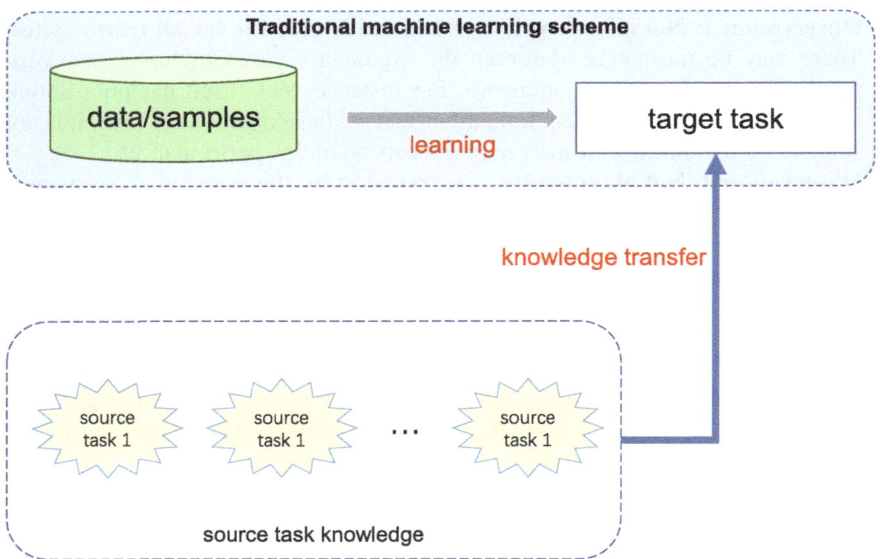

Fig. 5.1 The idea of transfer learning

be initiated by the results derived from the corresponding source task, as shown in Fig. 5.2. In the algorithm, the context-specific learning experience in terms of Q tables $Q_{(s,\mathcal{N}^*,i)}$ is stored in a database. Once it is found that there is already some learning record for current context $(s, \mathcal{N}^*, i)$, the learned Q table will be used for initiation. Otherwise, the Q table is initiated with a 0 vector, as shown in the first to fifth line of the algorithm. Accordingly, the initial exploration probability is $\varepsilon' < \varepsilon''$. In the loop, the algorithm updates the Q table and also detects the context change. Once the context varies due to the change of traffic type, available network set, or time period, the learned experience in terms of the Q table will be stored and followed by a restart of the algorithm with the new context $(\widehat{s}, \widehat{\mathcal{N}}^*, \widehat{i})$. This process realizes the context-dependent learning and knowledge transfer.

We make several remarks on the algorithm. First, the introduction of knowledge transfer mainly modifies the Q table according to contexts and has no change to the learning framework, thus, the convergence of the transferred reinforcement learning still holds [13]. Second, there is a concern about the division of time periods in the algorithm. Given the fixed traffic type location variation pattern, the resolution of time periods affects the performance and convergence of the proposed algorithm. Apparently, a larger time period length indicates a smaller context vector space and a longer learning experience length T for each context vector. However, this may experience varying load statistical distributions and thus incur negative learning experiences. A shorter time period length could provide more fine-grained contextual differentiation and a larger context vector space, which indirectly reduces the sample number in reusing experience for each context vector. Therefore, the division of time period

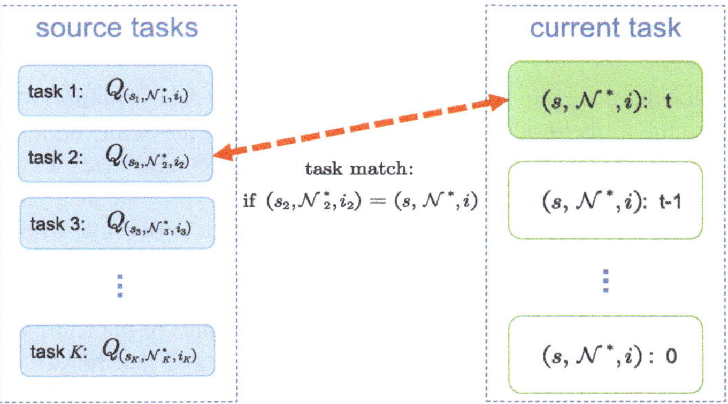

Fig. 5.2 The transfer learning in the proposed algorithm. The source task and the target task are mapped according to the context vector

should be carefully evaluated according to the evolution law of network load statistical distributions. Third, although the size of saved Q tables in the database will grow linearly as the increase of the experienced contexts or situations, the storage complexity of the algorithm is very limited due to: the number of typical context is small, e.g., home, office, and playground, and the stored information (context vector and Q table) is very limited for each context. Fourth, thanks to the learning experience reuse, the proposed algorithm can greatly cut down the exploration frequency, that is, the visit of random selection behavior in ninth line of the algorithm is restrained. Thus, the effect of "ping-pong" and associated handoff cost is greatly reduced, compared with the standard reinforcement learning. The negative effects can be further alleviated by carefully choosing the slot duration and resorting to some multipath concurrent transmission protocols such as the stream control transmission protocol that is able to provide multi-homing and redundant paths facilitating smooth network handoff with low cost. Finally, the context vector can be easily extended to include other factors if a finer context resolution is needed, such as user's activity description, age, preference, and some other user profiles. In addition, different users may share their learning experience to further improve the learning efficiency if they have common context vectors.

5.4   Simulation Results

In this section, we analyze the performance of the proposed algorithm.

5.4.1 Simulation Setting

We assume that there are one LTE small cell, two WLAN access points, and one VLC access point. Due to the modulation and coding scheme, the achieved instantaneous rate is discrete, which is determined by the user's location and varies with the fading effect over time. Following a similar idea in reference [14], we make a set of discrete achievable peak rates $R_{1,k} < R_{2,k} < \cdots < R_{M_k,k}$ for each network k, where M_k is the maximum number of achievable rates in network k. The data rates of the LTE small cell and WLAN are set by referring to some measured data from the "Speedtest" app. Specifically, the dynamic ranges of downlink date rates of the LTE, WLAN, and VLC are [4000 kbps, 7000 kbps], [3000 kbps, 10000 kbps] and [8000 kbps, 13000 kbps], and the uplink date rates are [500 kbps, 6000 kbps], [3000 kbps, 9000 kbps], and [80 kbps, 120 kbps], respectively. We adjust the actual numbers of active users and the data rate dynamic ranges of the four networks to create network performance diversity.

The utility function related parameters are set as $\Theta_0 = 50$ kbps, $\Theta_1 = 100$ kbps, $\Theta_2 = 8000$ kbps, $\lambda = 1$, $c = 10$, and $\lambda = 1$. The algorithm related parameters are set as $\gamma = 0.3$, $\varepsilon' = 0.1$ or 0.3, $\varepsilon'' = 0.3$.

5.4.2 Results

In the first group, we run the proposed algorithm to assess its convergence and compare it with standard Q-learning in a given context. The maximal number of active users in each network is 6. Note that since the uplink of VLC could hardly support the high uplink performance requirement, we remove VLC to reduce the action set for uplink-dominated traffic. The Q-learning with knowledge transfer used the learnt Q values averaged at the 100th slot of standard Q-learning algorithm. Figures 5.3, 5.4 and 5.5 show the performance comparison of the two algorithms. We can observe that the proposed Q-learning with knowledge transfer algorithm converges much faster and achieves larger average reward than that of the standard Q-learning algorithm. Specifically, for the downlink-dominated and uplink–downlink symmetric traffic types, the proposed algorithm nearly achieves convergence nearly at the first ten iterations. However, for uplink-dominated traffic, the performance gain does not appear before approximately 150 iterations. We infer that due to slow convergence speed, the experience at 100th iteration misled the algorithm. To verify this speculation, we checked the Q values at 100th and 200th iterations and found that the Q values at 100th did not match the converged one. This reminds us of the importance of sufficient experience accumulation. To show this point, we compared the performance with different Q value experience for the proposed algorithm. As shown in Fig. 5.6, the Q values at 50th and 100th slots does not help the convergence at the beginning of the proposed algorithm and even leads to poor performance before

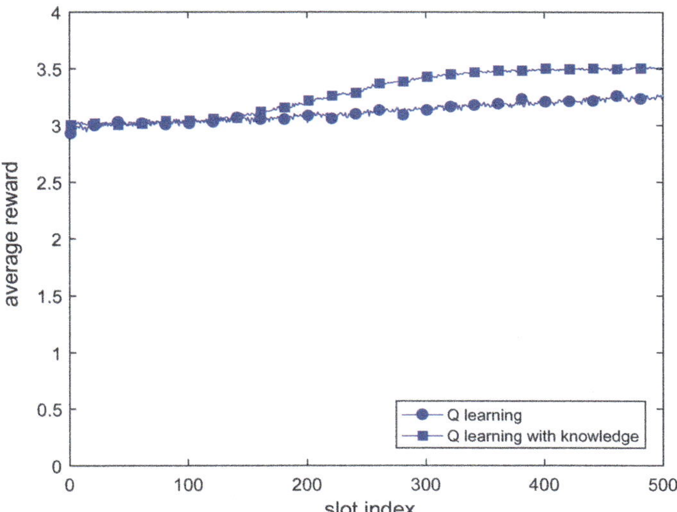

Fig. 5.3 Performance comparison with uplink-dominated traffic

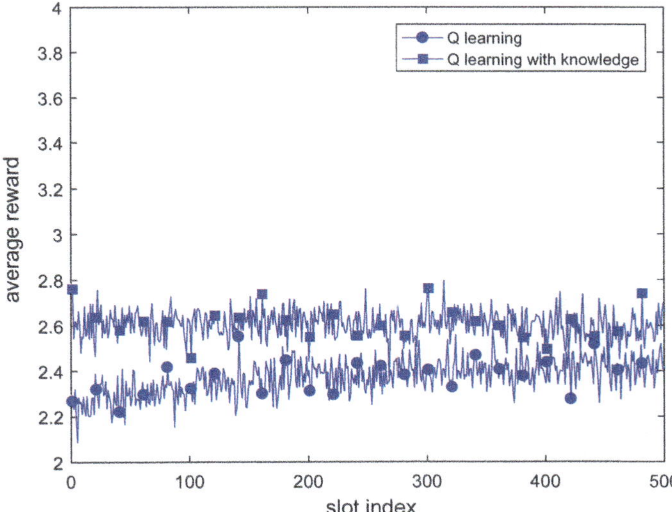

Fig. 5.4 Performance comparison with downlink-dominated traffic

200th iteration. On the other hand, converged Q values experience such as that at 200th iteration could improve the performance at the beginning stage.

Figures 5.7 and 5.8 show the related results with different exploration probabilities in the case of context evolution. It is assumed that the traffic type s in $(s, \mathcal{N}^*, i)$

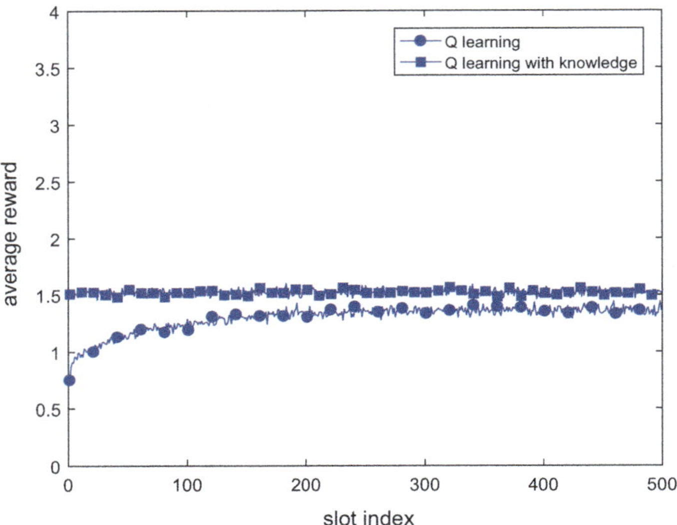

Fig. 5.5 Performance comparison with uplink–downlink symmetric traffic

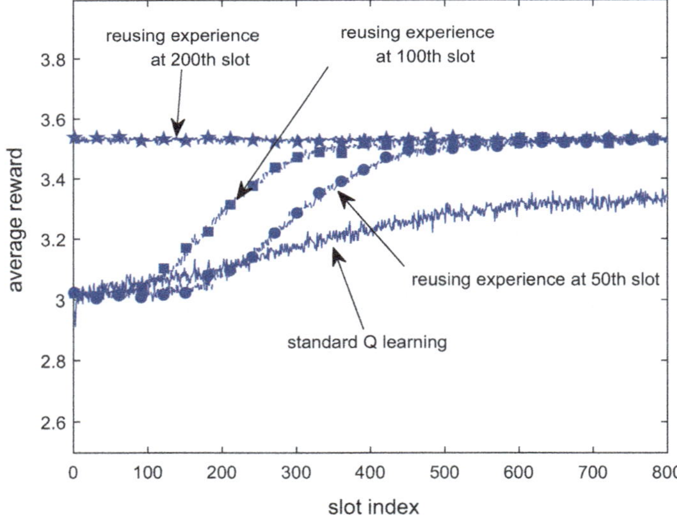

Fig. 5.6 Algorithm performance with different Q value experience

evolves as "uplink-dominated $\rightarrow$ downlink-dominated $\rightarrow$ uplink–downlink symmetric $\rightarrow$ uplink-dominated $\rightarrow$ downlink-dominated $\rightarrow$ uplink–downlink symmetric" and each traffic type lasts for 50 slots. Moreover, we generate different performances (data rates and user number distributions) of all networks for the the first 150 slots

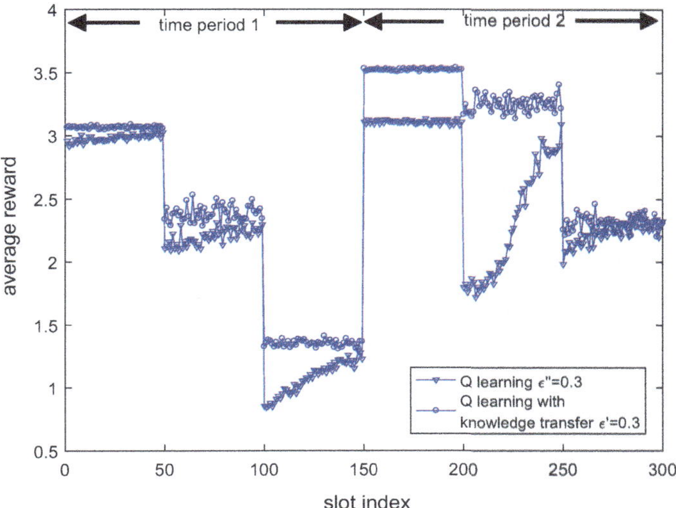

Fig. 5.7 Algorithm performance with $\epsilon = 0.3$

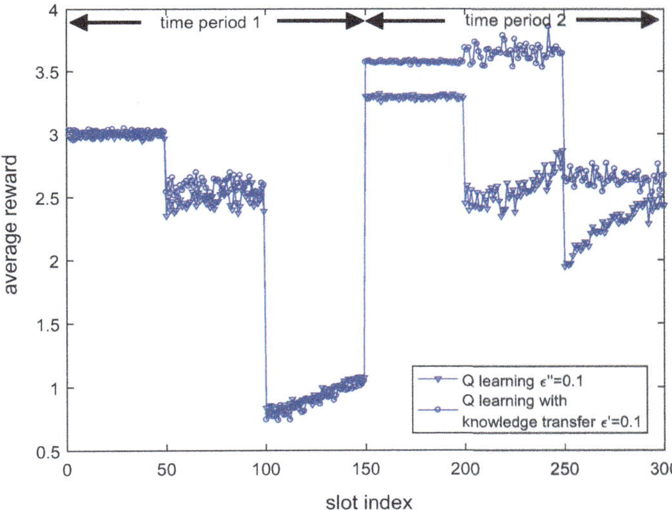

Fig. 5.8 Algorithm performance with $\epsilon = 0.1$

and the second 150 slots. We can see in both cases, the proposed algorithm performs better than the standard Q-learning algorithm in most stages. However, when the exploration probabilities is 0.1, the two algorithms seem to have the same performance during the first time period (1st to 50th slots and 100th to 150th slots). The reason is similar to Fig. 5.6, that is, the slow convergence constrained its performance improvement.

5.5 Conclusion

This chapter studied context-aware network selection. The problem is formulated as an MDP, where a context vector is introduced to enable context-dependent network selection. To improve the convergence of learning algorithm, we exploit the features of context information to design a Q-learning with knowledge transfer. The advantages of the proposed algorithm lie on two aspects: prior knowledge is used to reduce the action set and context-based learning experience in terms of Q values is reused to speed up the learning process.

References

1. Sim GH, Klos S et al (2018) An online context-aware machine learning algorithm for 5G mmWave vehicular communications. IEEE ACM T Network 26(6):2487–2500
2. Pantisano F, Bennis M, Saad W et al (2013) Matching with externalities for context-aware user-cell association in small cell networks. In: IEEE global telecommunications conference (GLOBECOM)
3. Yang H, Du P et al (2019) Reinforcement learning-based intelligent resource allocation for integrated VLCP systems. IEEE Wirel Commun Lett 8(4):1204–1207
4. Monteiro A, Souto E et al (2019) Context-aware network selection in heterogeneous wireless networks. Comput Commun 135:1–15
5. Du Z, Wang C, Sun Y, Wu G (2018) Context-aware indoor VLC/RF heterogeneous network selection: reinforcement learning with knowledge transfer. IEEE Access 6:33275–33284
6. Du Z, Wu Q, Yang P et al (2015) Exploiting user demand diversity in heterogeneous wireless networks. IEEE Trans Wirel Commun 14(8):4142–4155
7. Bianchi G (2000) Performance analysis of the IEEE 802.11 distributed coordination function. IEEE J Sel Areas Commun 8(3):535–547
8. Basnayaka DA Haas H (2015) Hybrid RF and VLC systems: improving user data rate performance of VLC systems. In: IEEE Vehicular Technology Conference (VTC)
9. Kavehrad M (2010) Sustainable energy-efficient wireless applications using light. IEEE Commun Mag 48(12):66–73
10. GigaIR, Infrared Data Association Standards, http://irda.org
11. González MC, Hidalgo CA, Barabási AL (2008) Understanding individual human mobility patterns. Nature 453:779–782
12. Lee D, Zhou S, Zhong X et al (2014) Spatial modeling of the traffic density in cellular networks. IEEE Wirel Commun 21(1):80–88
13. Talvitie E, Singh S (2007) An experts algorithm for transfer learning. In: international joint conference on artificial intelligence (IJCAI)
14. Ibrahim M, Khawam K, Tohme S (2010) Congestion games for distributed radio access selection in broadband networks. In: IEEE global telecommunications conference (GLOBECOM)

Part III
Game-Theoretic MARL Online Network Selection

This part explicitly studies multiple users' network selection behavior. The challenge of multiple users scenario lies in the interaction of individual users' decision-making, leading to a nonstationary environment and most importantly, low system efficiency due to users' competition. In addition, this part considers the heterogeneous user demand feature, which makes the problem even complicate. This part adopts the game theory to analyze the network selection problem, where the focus is to design an efficient multi-agent reinforcement learning (MARL) algorithm to solve it. Specifically, a localized cooperation game and a QoE game and associated MARL algorithms are proposed in Chap. 6 and Chap. 7, respectively.

Chapter 6
Matching Heterogeneous User Demands: Localized Self-organization Game and MARL Based Network Selection

Abstract This chapter focuses on network selection for multiple user cases. Since users' network selection decision determines the load distribution of networks, users' decision-making is interacted. In particular, when heterogeneous user demand is considered, the solution of the optimal match between users and networks becomes a challenge. Centralized solutions could achieve a fair performance at a high optimization cost. Distributed solutions incur less cost but commonly result in low efficiency due to user competition. Different from centralized approaches or distributed approaches, we propose a local improvement algorithm, where networks that share users, called coupled network pairs (CNPs), cooperatively re-associate users with user demand awareness. Under a novel localized self-organization game formulation, we proved that the local improvement algorithm can achieve promising performance. To speed up the convergence of the algorithm, we further exploit the spatial independence among CNPs and propose an enhanced local improvement algorithm. Finally, simulation results indicate that the proposed algorithms achieve much better performance with relatively short convergence time, compared with three distributed algorithms.

6.1 Introduction

Users' transmission performance in networks is affected by not only the physical channel quality, but also the load of the associated network. In essence, due to limited bandwidth of NAP, a higher load, i.e., the number of associated users, indicates that a less allocated resource or the resulting achieved throughput for individual users in the NAP. As a result, users' network selection decision-making is coupled: each individual user tries to maximize his own transmission performance while considering others' access behaviors. From the system perspective, the core of the multiple users network selection is to match users and NAP to maximize the global utility. In particular, when heterogeneous user demand is considered, the global optimization problem is challenging and even intractable due to the resulting combinatorial problems.

© Springer Nature Singapore Pte Ltd. 2020

Z. Du et al., *Towards User-Centric Intelligent Network Selection in 5G Heterogeneous Wireless Networks*,

https://doi.org/10.1007/978-981-15-1120-2_6

There are two types of methods to solve the global optimization in the existing literature. One typical method is centralized optimization. In related works [1, 2], some centralized controller is assumed to be responsible for the global optimization. Commonly, centralized optimization methods could achieve fair performance at the price of high optimization cost in terms of infrastructure supporting, information and controlling message exchange, high computation complexity, and so on. On the other hand, distributed methods distribute decision-making to individual users who decide their NAP based on local information. Many existing game-based approaches [3–5] adopt Nash equilibrium (NE) for the solution of distributed network selection. While distributed methods require less optimization cost, the system efficiency is relatively low due to user competition. Notably, the optimization cost issue in centralized methods and efficiency issue in distributed methods become even worse as the number of user terminals in 5G and beyond network grows fast.

To balance the optimization cost and optimization efficiency, similar to the hybrid optimization method in [6], we propose a local self-organizing network (L-SON) architecture, where NAPs with overlapping coverage are formed as optimization units to cooperatively re-associate users with user demand awareness. With this idea, we model the network selection as a localized self-organization game and prove the existence of NE and its optimality, i.e., the best NE corresponds to the optimal solution of the system efficiency. Finally, we propose a local improvement algorithm and its enhanced version to effectively achieve NE. The main results of this chapter were presented in [7].

6.2 System Model and Problem Formulation

6.2.1 System Model

Consider a given area, there are N partially overlapping NAPs $\mathcal{N} = \{n_1, n_2, \ldots, n_N\}$. A set of M users $\mathcal{M} = \{1, 2, \ldots, M\}$ are distributed in the coverage area of NAPs, as illustrated in Fig. 6.1. The focus is the network selection of users in overlapping areas, since they have multiple networks available and can access any one of them.

Denote the set of users associated with network n as $\mathcal{M}_n \in \Psi_n$, where Ψ_n is the set of all possible $\mathcal{M}_n$. We refer the vector of user set satisfying the condition $\bigcap_{n \in \mathcal{N}} \mathcal{M}_n = 0$, $\bigcup_{n \in \mathcal{N}} \mathcal{M}_n = \mathcal{M}$ as joint user–network association $\mathbf{M}_0 = [\mathcal{M}_1, \mathcal{M}_2, \ldots, \mathcal{M}_M]$, $\mathbf{M}_0 \in \Omega$, where Ω is the set space of $\mathbf{M}_0$. Note that the above conditions on $\mathbf{M}_0$ ensures that each user is associated with one and only one network. Let $n(m) \in \mathcal{N}$ be the network selected by user m. Given the limited network resource, the achieved throughput of user m, θ_m, is denoted by $\theta_m(\mathcal{M}_{n(m)}, \Pi)$, where Π is the resource schedule policy in networks. Explicitly, on one hand, when the number of users selecting n increases, i.e., $|\mathcal{M}_n|$ increases, the achieved throughput for individual user $m \in \mathcal{M}_n$ decreases. On the other hand, the resource schedule policy

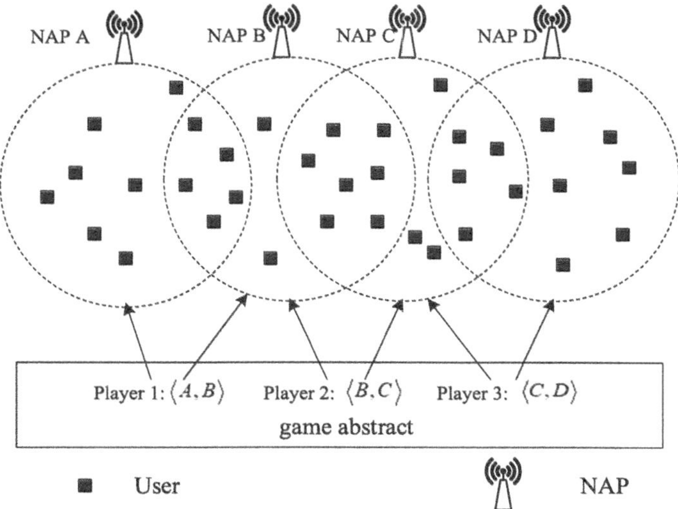

Fig. 6.1 System model and abstracted localized self-organization game

Π determines the exact resource schedule for individual user $m \in \mathcal{M}_n$. The resource schedule policy Π is assumed fixed. Therefore, only $\mathcal{M}_n$ determines the throughput for each user $m \in \mathcal{M}_n$.

6.2.2 Problem Formulation

Given the achieved throughput θ_m, user m's utility is $u_m(\theta_m)$, where $u_m(\cdot)$ is the user's demand function mapping throughput θ_m to user's utility. Considering the realistic scenario that users may have different service requirements, preference, and so on, the user demands can be diverse. Thus, the demand functions can be heterogeneous and user specific, which is more practical than most existing works. The goal is to find a joint user–network association $\mathbf{M}_0^*$ maximizing the utilization of network resource, i.e., maximizing the social welfare $U_{social}(\mathbf{M}_0)$ in terms of the total utility of users,

$$U_{social}(\mathbf{M}_0) = \sum_{m \in \mathcal{M}} u_m\left(\theta_m\left(\mathcal{M}_{n(m)}, \Pi\right).\right) \tag{6.1}$$

Note that users who locate in the overlapping area of networks can change their network selections. Their selection decisions directly determine $\mathbf{M}_0$. Users who locate in the areas covered by only one network, we call regular users, cannot change their access network. However, due to the resource sharing in a network, the regular users distribution will affect the optimal $\mathbf{M}_0^*$.

Although the considered model is general, we give a more concrete scenario with specific resource schedule policy and user utility functions for clarity. We differentiate users into three types according to user demands: brittle traffic user c_b, stream traffic user c_s, and elastic traffic user c_e (Different definitions or classifications on the user demand can also be used). The brittle traffic user represents users who need real-time traffic with strict performance requirements. For stream traffic users, they may want to watch some online multimedia, thus they also have some performance requirement, which is looser than the brittle traffic user. The elastic traffic user represents users who need non-real-time traffics such as E-mail or file transfer. They are not sensitive to the variation of throughput and don't have a minimal throughput requirement. Such kind of classification captures the throughput requirement characteristic of users with different user demands. The detailed information on the classification can refer to the literature [8]. Here, a service differentiation resource schedule policy as in [9] is adopted. For user m associated with network n, the achieved throughput is determined by

$$\theta_m = \frac{R_{m,n} w_m}{W_n}, \tag{6.2}$$

where $R_{m,n}$ is the peak data rate of the link between user m and network n, w_m is user m's weight, $W_n = \sum_{i \in M_n} w_i$ is the total weight of users associated with network n. The user weights can be different, which depend on the specific user type. Note that although WLAN does not offer service differentiation with such proportional fairness, there exist several solutions to address this issue [9]. The weights for each brittle traffic user, stream traffic user, and elastic traffic user are w_b, w_s, and w_e, respectively. Denote $\mathbf{W} = [w_b, w_s, w_e]$ as the corresponding weight vector.

When the user demand function is involved, the utility function can be expressed as

$$u_m \left(\theta_m, c_m \right) = \begin{cases} f \left(\theta_m \right), & c_m = c_b \\ g \left(\theta_m \right), & c_m = c_s \\ h \left(\theta_m \right), & c_m = c_e, \end{cases}$$

where c_m is the user type, $f \left(\theta \right)$, $g \left(\theta \right)$, and $h \left(\theta \right)$ are the user utility functions for brittle traffic user, partially elastic traffic user, and elastic traffic user, respectively.

The optimal user–network association depends on the user number as well as user type distribution in networks. Denote the achieved throughput of user m given joint user–network association $\mathbf{M}_0$ as $\theta_m \left(\mathcal{M}_{n(m)} \right)$, the problem can be formally formulated as

$$\underset{\mathbf{M}_0 \in \Omega}{\text{maximize}} \, U_{social} \left(\mathbf{M}_0 \right) = \sum_{m \in M} u_m \left(\theta_m \left(\mathcal{M}_{n(m)} \right), c_m \right). \tag{6.3}$$

We emphasize that the utility function $u_m \left(\theta_m \right)$ intends to characterize diverse user demands in reality.

6.3 Local Self-organizing Network-Based Network Selection

In this section, we introduce the optimization architecture of L-SON and propose a localized self-organizing game to model the network selection.

6.3.1 The Idea of L-SON

In the most general case, solving **P**.1 is, in fact, to optimally associate users in the overlapping areas with networks, which falls within the combinatorial optimization problem that proved to be NP-hard [10]. Obviously, its complexity grows as the number of users in the overlapping areas increases. As we have mentioned above, traditional centralized optimization methods require high optimization cost for global information and signaling exchange, and distributed optimization methods will lead to low system efficiency due to user competition. In order to derive practical solutions, we attempt to design a new approach.

Actually, we can see that the spatial distribution of networks generally forms finite overlapping areas as shown in Fig. 6.1. The user–network associations in overlapping areas jointly determine the social welfare. Optimizing the user–network association in a single overlapping area is much easier than optimizing the global objective in (6.3). Can we approximate the global objective on the basis of optimization in individual overlapping areas? Generally, due to the complex interactions among different overlapping areas, optimizing the social welfare is not equivalent to directly optimizing the user network association of each overlapping area. Hence, in order to have fair approximation on the basis of optimization in individual overlapping area, it calls for additional regulation and design. Hence, we propose the idea of L-SON: NAPs with (partial) overlapping coverage cooperatively re-associate the user–network association for users in the overlapping area to achieve maximization of the global objective in (6.3).

6.3.2 Localized Self-organization Game Formulation

Before formally introducing the game model, some notations and variables are given. We use $\mathbf{D}_{\langle n_i,n_j \rangle} = \left[\alpha_{\langle n_i,n_j \rangle}, \beta_{\langle n_i,n_j \rangle}, \chi_{\langle n_i,n_j \rangle} \right]$ to represent the fundamental user distribution in the overlapping area of network n_i and n_j, where $\alpha_{\langle n_i,n_j \rangle}$, $\beta_{\langle n_i,n_j \rangle}$, $\chi_{\langle n_i,n_j \rangle}$

are the numbers of brittle traffic users, stream traffic users, and elastic traffic users in the overlapping area. For example, $\mathbf{D}_{\langle n_i,n_j\rangle} \neq \mathbf{0}$ indicates that there is at least one user located in the overlapping area of network n_i and n_j, while $\mathbf{D}_{\langle n_i,n_j\rangle} = \mathbf{0}$ indicates that there is no user located in the overlapping area. Note that three types of users are differentiated in the fundamental user distribution.

Then, we give definitions of the coupled network pair (CNP), the user distribution of networks, the user distribution of CNPs, and the joint user distribution of CNPs, respectively. We define any network pair n_i and n_j satisfying $\mathbf{D}_{\langle n_i,n_j\rangle} \neq \mathbf{0}$ as a CNP and denote it as $\langle n_i, n_j\rangle$. $\langle n_i, n_j\rangle$ and $\langle n_j, n_i\rangle$ are not differentiated and are treated as only one CNP. The corresponding CNP set in the whole system is C. Given any CNP $\langle n_i, n_j\rangle$, we denote $\mathbf{x}^{n_i}_{\langle n_i,n_j\rangle} = \left[\alpha^{n_i}_{\langle n_i,n_j\rangle}, \beta^{n_i}_{\langle n_i,n_j\rangle}, \chi^{n_i}_{\langle n_i,n_j\rangle}\right]$ as the user distribution of network n_i in CNP $\langle n_i, n_j\rangle$, where $\alpha^{n_i}_{\langle n_i,n_j\rangle} \geq 0, \beta^{n_i}_{\langle n_i,n_j\rangle} \geq 0, \chi^{n_i}_{\langle n_i,n_j\rangle} \geq 0$ are the numbers of brittle traffic users, stream traffic users and elastic traffic users associated to network n_i in the overlapping area of CNP $\langle n_i, n_j\rangle$. Similarly, $\mathbf{x}^{n_j}_{\langle n_i,n_j\rangle}$ is the user distribution of network n_j in the overlapping area of CNP $\langle n_i, n_j\rangle$. We denote $\mathbf{x}_{\langle n_i,n_j\rangle} = \left[\mathbf{x}^{n_i}_{\langle n_i,n_j\rangle}, \mathbf{x}^{n_j}_{\langle n_i,n_j\rangle}\right]$ satisfying $\mathbf{x}^{n_i}_{\langle n_i,n_j\rangle} + \mathbf{x}^{n_j}_{\langle n_i,n_j\rangle} = \mathbf{D}_{\langle n_i,n_j\rangle}$ as the user distribution of CNP $\langle n_i, n_j\rangle$, which is an element of the possible user distribution set $\mathbf{X}_{\langle n_i,n_j\rangle}$. Obviously, $\mathbf{x}_{\langle n_i,n_j\rangle}$ is the actual user–network association in the overlapping area. $\mathbf{X}_{\langle n_i,n_j\rangle}$ is determined by $\mathbf{D}_{\langle n_i,n_j\rangle}$ whose cardinality is $\left|\mathbf{X}_{\langle n_i,n_j\rangle}\right| = \left(\alpha_{\langle n_i,n_j\rangle} + 1\right)\left(\beta_{\langle n_i,n_j\rangle} + 1\right)\left(\chi_{\langle n_i,n_j\rangle} + 1\right)$. For instance, given the number of brittle traffic users, stream traffic users, and elastic traffic users in the overlapping area of n_i and n_j as 4, 5, and 3, respectively, then the fundamental user distribution is $D_{<n_i,n_j>} = [3, 4, 5]$ and the number of possible user distribution set is $\left|\mathbf{X}_{\langle n_i,n_j\rangle}\right| = 5 \cdot 6 \cdot 4 = 120$. If 2 brittle traffic users, 3 stream traffic users, and 1 elastic traffic user of these users are currently associated with network n_i, then the user distributions can be represented as $x^{n_i}_{<n_i,n_j>} = [2, 3, 1]$ and $x^{n_j}_{<n_i,n_j>} = [2, 2, 2]$.

A joint user distribution $\mathbf{x} = (\mathbf{x}_1, \mathbf{x}_2, \ldots, \mathbf{x}_C)$ is a user distribution vector of all CNPs, the corresponding set of joint user distribution is $\mathbf{X} = \underset{\langle n_i n_j\rangle \in C}{\times} \mathbf{X}_{\langle n_i n_j\rangle}$, where $\times$ is the Cartesian product. In addition, $\mathbf{y}^{n_i} = [\alpha^{n_i}, \beta^{n_i}, \chi^{n_i}]$ is used to represent the regular user distribution in network n_i, where $\alpha^{n_i} \geq 0, \beta^{n_i} \geq 0, \chi^{n_i} \geq 0$ are the the numbers of brittle traffic users, stream traffic users, and elastic traffic users who can access the sole network n_i. $\mathbf{y} = \left[\mathbf{y}^1, \mathbf{y}^2, \ldots, \mathbf{y}^N\right]$ is the joint regular user distribution. Similar to [11], we assume that users with the same type have the same bit rate in a network, that is, $R_{m,n} = R_{m',n}, \forall m, m' \in \mathcal{M}, c_m = c_{m'}, \forall n \in \mathcal{N}$. In this context, $R_{m,n}$ can be explained as the bandwidth of network n, the problem is reduced to determine the user number of each user type associated with each network, thus $\mathbf{x}^{n_i}_{\langle n_i,n_j\rangle}$ can fully characterize the user–network association in the overlapping area of n_i and n_j. This assumption simplifies our representation, while our algorithms can be extended for the general cases where users have specific date rates $R_{m,n}$.

Now, we can formulate a user–network association game by referring to each CNP $\langle n_i, n_j \rangle \in C$ as a player, whose strategy is the user distribution $\mathbf{x}_{\langle n_i, n_j \rangle}$. The pure strategy set of each player $\langle n_i, n_j \rangle$ is the user distribution set $\mathbf{X}_{\langle n_i, n_j \rangle}$. The strategy profile of all players is the joint user distribution $\mathbf{x}$ and the set of strategy profile is $\mathbf{X}$. Each player $\langle n_i, n_j \rangle$ selects a strategy $\mathbf{x}_{\langle n_i, n_j \rangle} \in \mathbf{X}_{\langle n_i, n_j \rangle}$ to maximize its utility defined by

$$U_{\langle n_i, n_j \rangle}(\mathbf{x}) = \sum_{m \in \mathcal{M}_{n_i}(\mathbf{x})} u_m(\theta_m, c_m) + \sum_{m \in \mathcal{M}_{n_j}(\mathbf{x})} u_m(\theta_m, c_m), \tag{6.4}$$

where $\mathcal{M}_{n_i}(\mathbf{x})$ and $\mathcal{M}_{n_j}(\mathbf{x})$ are the set of users associated with network n_i and n_j given user distribution $\mathbf{x}$, respectively. $\theta_m\left(\mathcal{M}_{n(m)}\right)$ is denoted as θ_m for simplicity from now on. We denote the user–network association game as

$$\Gamma = \left(C, \left(\mathbf{X}_{\langle n_i, n_j \rangle} \right)_{\langle n_i, n_j \rangle \in C}, \left(U_{\langle n_i, n_j \rangle} \right)_{\langle n_i, n_j \rangle \in C} \right)$$

For example, we can abstract a user–network association game of the scenario in Fig. 6.1 where three players exist: $\langle A, B \rangle$, $\langle B, C \rangle$, and $\langle C, D \rangle$.

In the following, we give the definition of Nash equilibrium.

Definition 6.1 A strategy profile $\mathbf{x}$ is a pure strategy Nash equilibrium if no CNP can improve its utility by unilaterally changing its strategy, i.e.,

$$U_{\langle n_i, n_j \rangle}(\mathbf{x}) \geq U_{\langle n_i, n_j \rangle}\left(\mathbf{x}'_{\langle n_i, n_j \rangle}, \mathbf{x}_{-\langle n_i, n_j \rangle} \right), \forall \langle n_i, n_j \rangle \in C, \forall \mathbf{x}'_{\langle n_i, n_j \rangle} \in \mathbf{X}_{\langle n_i, n_j \rangle}.$$

It is necessary to explain the relationship between the user–network association and the user distribution. Given a joint user–network association $\mathbf{M}_0$, the user distribution of any CNP $\langle n_i, n_j \rangle$ can be obtained as $\mathbf{x}_{\langle n_i, n_j \rangle}(\mathbf{M}_0)$. While given the joint user distribution of all CNPs $\mathbf{x}$, there is a corresponding set of possible $\mathbf{M}_0 \in \Omega(\mathbf{y}, \mathbf{x})$, since the user–network association differentiates user individuals but the user distribution only differentiates user types. Nevertheless, the social welfare of joint user–network associations in $\Omega(\mathbf{y}, \mathbf{x})$ are the same as the social welfare given $\mathbf{y}$ and $\mathbf{x}$, i.e.,

$$U_{social}\big|_{(\mathbf{y}, \mathbf{x})} = U_{social}(\mathbf{M}_0), \forall \mathbf{M}_0 \in \Omega(\mathbf{y}, \mathbf{x}). \tag{6.5}$$

Considering that $\mathbf{y}$ is fixed, each $\mathbf{x}$ is sufficiently to determine a specific social welfare. Therefore, we can assume that $\mathbf{M}_0(\mathbf{x})$ corresponds to only one of joint user–network associations in $\Omega(\mathbf{y}, \mathbf{x})$ for clarity.

6.4 Algorithm Design

On the basis of game formulation, this section proposes two simple but effective algorithms. In particular, we can prove that the two algorithms converge to pure NE of the localized self-organization game with a performance guarantee.

6.4.1 Local Improvement Algorithm

The process of LIA is shown in Algorithm 1. Let $\mathbf{x}_{-\langle n_i,n_j\rangle}$ represent the user distribution of all CNPs except CNP $\langle n_i, n_j\rangle$, we can denote $\mathbf{x}$ as $\mathbf{x} = \left(\mathbf{x}_{\langle n_i,n_j\rangle}, \mathbf{x}_{-\langle n_i,n_j\rangle}\right)$ with some abuse of notation. Starting from any initial user distribution $\mathbf{x}^0_{\langle n_i,n_j\rangle}$ of each CNP, the algorithm randomly chooses one CNP to update in each iteration. The two networks in the selected CNP re-associate the shared users in the overlapping area as $\mathbf{x}'_{\langle n_i,n_j\rangle} \in \mathbf{X}_{\langle n_i,n_j\rangle}$ to maximize the total utility of users currently associated with the two networks, $U_{\langle n_i,n_j\rangle}\left(\mathbf{x}_{\langle n_i,n_j\rangle}, \mathbf{x}_{-\langle n_i,n_j\rangle}\right)$. Note that the above process only needs local information exchange in the CNP. Actually, in each iteration, only one CNP is updated in LIA, which ensures that neighboring CNPs keep the user distributions unchanged. Given the fixed $x_{-<n_i,n_j>}$ (regardless its specific value), it is sufficient for CNP $< n_i, n_j >$ to make the above decision by selecting its own user distributions among networks. Take the running of LIA on Fig. 6.1 as an example, there are three overlapping areas corresponding to three CNPs $\langle A, B\rangle$, $\langle B, C\rangle$, $\langle C, D\rangle$. In each iteration, one pair of network is activated to re-associate users in overlapping area to maximize the total utility of users in these two networks.

Algorithm 8 Local improvement algorithm: LIA

1: **Initiate:** Given some initiated $\mathbf{x}^0_{\langle n_i,n_j\rangle}$.
2: **loop**
3:    A random CNP $\langle n_i, n_j\rangle \in C$ is activated.
4:    The netwroks in the CNP n_i, n_j reallocate the users in the overlapping area of them as $\mathbf{x}'_{\langle n_i,n_j\rangle}$
      to maximize the total utilities of users in these two networks,

$$\mathbf{x}'_{\langle n_i,n_j\rangle} = \arg \max_{\mathbf{x}_{\langle n_i,n_j\rangle} \in \mathbf{X}_{\langle n_i,n_j\rangle}} U_{\langle n_i,n_j\rangle}\left(\mathbf{x}_{\langle n_i,n_j\rangle}, \mathbf{x}_{-\langle n_i,n_j\rangle}\right)$$

5: **end loop**

LIA decomposes the complex optimization problem into multiple subproblems easy to handle distributively. Each subproblem is solved under the localized cooperation of networks in a CNP. Note that the information exchange and coordination among networks are feasible, since there are several standards supporting information exchange among heterogeneous networks. For example, the LTE-A supports the information exchange between macrocells and femtocells, the IEEE 802.21 and some new emerging standards in [12] also support information exchanges among heterogeneous networks. Depending on the specific scenarios, finding $\mathbf{x}'_{\langle n_i,n_j\rangle}$ may resort to different tools such as dynamic programming. In the most general scenarios, an exhaust search in $\mathbf{X}_{\langle n_i,n_j\rangle}$ can be used to get the updated $\mathbf{x}'_{\langle n_i,n_j\rangle}$. Indeed, the exhaust search is relatively light weight considering the limited number of users. Despite its simplicity, LIA can be proved to converge to a equilibrium state that each CNP in C will maintain a user distribution $\mathbf{x}_{\langle n_i,n_j\rangle}$. The corresponding result is shown in Theorem 6.1.

Theorem 6.1 *LIA converges to a local or global optimum of the social welfare in 6.3. Specially, the global optimum social welfare is one of the equilibriums in LIA.*

Proof The proof relies on the feature of ordinal potential game as follows.

Lemma 6.1 *Every finite ordinal potential game has at least one pure strategy Nash equilibrium, which is a local or global maximum point of the potential function.*

We can prove that the game Γ belongs to ordinal potential game in the following.

Lemma 6.2 *The game Γ is a finite exact potential game and the social welfare $U_{social}\,(\mathbf{M}_0)$ is the potential function.*

Proof Define a function $\Phi : \mathbf{X} \to \mathfrak{R}$ as $\Phi\,(\mathbf{x}) = \sum_{n\in\mathcal{N}}\sum_{m\in\mathcal{M}_n(\mathbf{x})} u_m\,(\theta_m, c_m)$. Note that the strategy change of a CNP $\langle n_i, n_j\rangle$ from $\mathbf{x}_{\langle n_i,n_j\rangle}$ to $\tilde{\mathbf{x}}_{\langle n_i,n_j\rangle}$ has no influence on $\sum_{m\in\mathcal{M}_n(\mathbf{X})} u_m\,(\theta_m, c_m)$ for $\forall n \in \mathcal{N}, n \neq n_i$ and $n \neq n_j$. Hence, the following Eq. (6.6) holds.

$$
\mathcal{M}_n\left(\mathbf{x}_{\langle n_i,n_j\rangle}, \mathbf{x}_{-\langle n_i,n_j\rangle}\right) = \mathcal{M}_n\left(\tilde{\mathbf{x}}_{\langle n_i,n_j\rangle}, \mathbf{x}_{-\langle n_i,n_j\rangle}\right)
$$

$$
= \mathcal{M}_n\left(\mathbf{x}_{-\langle n_i,n_j\rangle}\right), \qquad \forall n \in \mathcal{N}, n \neq n_i, n \neq n_j \qquad (6.6)
$$

Given all the other CNPs' strategies $\mathbf{x}_{-\langle n_i,n_j\rangle}$, the change in Φ by CNP $\langle n_i, n_j\rangle$ unilaterally deviating its strategy from $\mathbf{x}_{\langle n_i,n_j\rangle}$ to $\tilde{\mathbf{x}}_{\langle n_i,n_j\rangle}$ is obtained as (6.7) according to Eq. (6.6).

$$
\Phi\left(\mathbf{x}_{\langle n_i,n_j\rangle}, \mathbf{x}_{-\langle n_i,n_j\rangle}\right) - \Phi\left(\tilde{\mathbf{x}}_{\langle n_i,n_j\rangle}, \mathbf{x}_{-\langle n_i,n_j\rangle}\right)
$$

$$
= \sum_{m\in\mathcal{M}_{n_i}\left(\mathbf{x}_{\langle n_i,n_j\rangle},\mathbf{x}_{-\langle n_i,n_j\rangle}\right)\cup\mathcal{M}_{n_j}\left(\mathbf{x}_{\langle n_i,n_j\rangle},\mathbf{x}_{-\langle n_i,n_j\rangle}\right)} u_m\,(\theta_m, c_m)
$$

$$
+ \sum_{n\in\mathcal{N},n\neq n_j,n\neq n_j}\;\sum_{m\in\mathcal{M}_n\left(\mathbf{x}_{-\langle n_i,n_j\rangle}\right)} u_m\,(\theta_m, c_m)
$$

$$
-\left[\sum_{m \in \mathcal{M}_{n_i}\left(\tilde{\mathbf{x}}_{\langle n_i,n_j \rangle}, \mathbf{x}_{-\langle n_i,n_j \rangle}\right) \cup \mathcal{M}_{n_j}\left(\tilde{\mathbf{x}}_{\langle n_i,n_j \rangle}, \mathbf{x}_{-\langle n_i,n_j \rangle}\right)} u_m\left(\theta_m, c_m\right) \right.
$$

$$
\left. + \sum_{n \in N, n \neq n_j, n \neq n_j} \sum_{m \in \mathcal{M}_n\left(\mathbf{x}_{-\langle n_i,n_j \rangle}\right)} u_m\left(\theta_m, c_m\right) \right]
$$

$$
= \sum_{m \in \mathcal{M}_{n_i}\left(\mathbf{x}_{\langle n_i,n_j \rangle}, \mathbf{x}_{-\langle n_i,n_j \rangle}\right) \cup \mathcal{M}_{n_j}\left(\mathbf{x}_{\langle n_i,n_j \rangle}, \mathbf{x}_{-\langle n_i,n_j \rangle}\right)} u_m\left(\theta_m, c_m\right)
$$

$$
- \sum_{m \in \mathcal{M}_{n_i}\left(\tilde{\mathbf{x}}_{\langle n_i,n_j \rangle}, \mathbf{x}_{-\langle n_i,n_j \rangle}\right) \cup \mathcal{M}_{n_j}\left(\tilde{\mathbf{x}}_{\langle n_i,n_j \rangle}, \mathbf{x}_{-\langle n_i,n_j \rangle}\right)} u_m\left(\theta_m, c_m\right). \tag{6.7}
$$

We found that the change in Φ equals to the change in $U_{\langle n_i,n_j \rangle}$, as shown in (6.8).

$$
\Phi\left(\mathbf{x}_{\langle n_i,n_j \rangle}, \mathbf{x}_{-\langle n_i,n_j \rangle}\right) - \Phi\left(\tilde{\mathbf{x}}_{\langle n_i,n_j \rangle}, \mathbf{x}_{-\langle n_i,n_j \rangle}\right)
$$

$$
= U_{\langle n_i,n_j \rangle}\left(\mathbf{x}_{\langle n_i,n_j \rangle}, \mathbf{x}_{-\langle n_i,n_j \rangle}\right) - U_{\langle n_i,n_j \rangle}\left(\tilde{\mathbf{x}}_{\langle n_i,n_j \rangle}, \mathbf{x}_{-\langle n_i,n_j \rangle}\right). \tag{6.8}
$$

According to the definition of potential game [13], the game Γ is an exact potential game with potential function Φ. Moreover, the joint strategy space is finite, thus, the game Γ is a finite exact potential game. It is worth noting that the potential function is exactly the social welfare, i.e., $\Phi(\mathbf{x}) = U_{social}(\mathbf{M}_0(\mathbf{x}))$. $\qquad \square$

Since any exact potential game is also an ordinal potential game, the existence of NE in the game Γ is guaranteed according to Lemmas 6.1 and 6.2. Because the potential function is also the social welfare, an NE of the user–network association game is a local or global optimum of the social welfare, where the best NE globally optimizes the social welfare. Note that the potential function increases in each iteration of LIA. Considering the finite states of the game, the potential function is bounded and LIA will finally terminate at some NE. This completes the proof. $\qquad \square$

Theorem 6.1 indicates that LIA will eventually converge to the global or local optimum equilibriums. Besides, we can also analyze its lower bound in Theorem 6.2.

Theorem 6.2 *The social welfare in LIA is larger than* $\dfrac{\sum_{\langle n_i,n_j \rangle \in C} \max\left\{U_{n_i}^{\max} + U_{n_j}^{\min}, U_{n_j}^{\max} + U_{n_i}^{\min}\right\}}{\max\limits_{n \in N} |\mathcal{H}_n|}$,
where $U_n^{\max}$ and $U_n^{\min}$ are two bounds of the total utility of users in network $n \in N$ (see in the Appendix), $\mathcal{H}_n = \left\{n' | \langle n, n' \rangle \in C\right\}$ is network n's neighbor set.

Proof The proof is omitted due to limited space. $\qquad \square$

6.4.2 Enhanced Local Improvement Algorithm

We found that the update of LIA can be more efficient in some cases. Consider a large-scale network distribution, there may exist spatial independence among CNPs, in the sense that they are not affected by each other's action. For instance, CNPs $\langle A, B \rangle$ and $\langle C, D \rangle$ are independent in Fig. 6.1, since they do not share NAPs.

Apparently, the spatial independence among CNPs indicates that multiple CNPs may update simultaneously. Motivated by this fact, we define the free CNP set C_f as any CNP set in which CNPs have no common networks with each other, i.e., $\mathcal{N}_c \cap \mathcal{N}_{c'} = \emptyset, \forall c, c' \in C_f, c \neq c'$, where $\mathcal{N}_c$ denotes the involved network set in c. Under this definition, there are four possible free CNP sets in Fig. 6.1: $\{\langle A, B \rangle\}$, $\{\langle B, C \rangle\}$, $\{\langle C, D \rangle\}$, $\{\langle A, B \rangle, \langle C, D \rangle\}$. We then propose an enhanced local improvement algorithm (E-LIA) which updates the user reassociation on a free CNP set basis in Algorithm 9. In E-LIA, a free CNP set C_f is randomly generated as follows: each network randomly selects one network in its neighbor set $\mathcal{H}_n$ and sends it the request to form a CNP, only CNPs whose networks are selected by each other is formed and activated. Then, the activated CNPs form the free CNP set C_f.

Algorithm 9 Enhanced local improvement algorithm: E-LIA

1: **Initiate:** Given some initiated $\mathbf{x}^0{}_{\langle n_i, n_j \rangle}$.

2: **loop**

3:     Each NAP randomly selects a neighboring NAP from the set $\mathcal{H}_n$ and send a CNP formulation request to it; Each NAP receives CNP formulation requests from other NAPs.

4:     Each NAP participate in at most one CNP. Activated CNPs are the free CNP set C_f in current slot.

5:     **loop**

6:         Denote the joint user distribution of all CNPs except $\langle n_i, n_j \rangle$ by $\mathbf{x}_{-\langle n_i, n_j \rangle}$. The two NAPs n_i and n_j reallocate the users in the overlapping area of them as $\mathbf{x}'{}_{\langle n_i, n_j \rangle}$ to maximize the total utilities of users in these two networks

$$\mathbf{x}'{}_{\langle n_i, n_j \rangle} = \arg \max_{\mathbf{x}_{\langle n_i, n_j \rangle} \in \mathbf{X}_{\langle n_i, n_j \rangle}} U_{\langle n_i, n_j \rangle} \left(\mathbf{x}_{\langle n_i, n_j \rangle}, \mathbf{x}_{-\langle n_i, n_j \rangle}. \right) \tag{6.9}$$

7:         Nonactivated CNPs maintain their current user association.

8:     **end loop**

9: **end loop**

The convergence performance of LIA and E-LIA can be briefly analyzed. Denote the set of possible free CNP sets as C_F, the maximal CNP number in a free CNP set is $T = \max_{C_f \in C_F} |C_f|$. Given any sample run of the LIA with t_1 iterations to converge, combing CNPs who can form a free CNP set in successive iterations and updating them simultaneously, we can get a corresponding sample run of the E-LIA with t_2 iterations to converge. The number of combined CNPs in one iteration of enhanced local improvement algorithm is at most T. Therefore, $\frac{t_1}{T} \leq t_2 \leq t_1$.

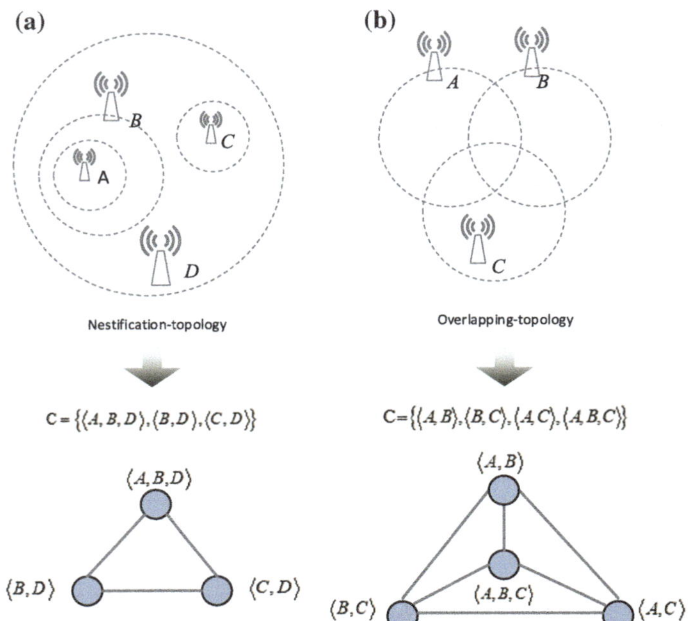

Fig. 6.2 CNP abstraction for different network topologies

Fig. 6.3 A four **a** vertex
claw graph and **b** claw-free
graph

Obviously, T is determined by the network distribution topology. To study T, we define a undirected CNP graph $G = \{C, \mathcal{E}\}$, where the CNP set is the vertex set, $\mathcal{E} \subset C \subseteq C$ is the edge set. Any two CNPs are connected by an edge if and only if they have at least one common network, i.e., $(c, c') \in \mathcal{E}$ if and only if $\mathcal{N}_c \cap \mathcal{N}_{c'} \neq \emptyset$. We also show the CNP graphs of different network distributions in Fig. 6.2. In the context of CNP graph, a free CNP set corresponds to an independent vertex set of G, and T is the size of the maximal independent vertex set. Because CNP graphs derive from realistic network distributions, its topology possesses distinct properties. Before progressing to the property of CNP graphs, we give some concepts in graph theory [14]. The degree $d(c)$ of vertex c is the number of edges at c. A graph is claw-free if no vertex has three pairwise nonadjacent neighbors. Figure 6.3 shows the comparison of four vertex claw graphs and claw-free graphs. The CNP graph has the following property.

Theorem 6.3 *CNP graphs are connected claw-free graphs.*

Proof We prove it by contradiction. Suppose a CNP graph $G = \{C, \mathcal{E}\}$ has claws. Then, there must exist at least one CNP c with $K (K \geq 3)$ neighboring CNPs $c_k, k = 1, 2, \ldots, K$, which are nonadjacent with each other. For each c_k, there exists at least one network $n_k \in \mathcal{N}_{c_k}$ satisfying $n_k \in \mathcal{N}_c \cap \mathcal{N}_{c_k}$. Since any two c_k are non-adjacent, i.e., $\bigcap_{k=1}^{K} \mathcal{N}_{c_k} = \emptyset$. Therefore, $n_{k1} \neq n_{k2}$, for $k1 \neq k2, 1 \leq k1, k2 \leq K$, which indicates that CNP c consists of at least K different wireless networks. This conclusion results in a network distribution where the area overlapped simultaneously by $K \geq 3$ networks is the only overlapping area of these networks. However, such kind of network distribution is impossible in reality.   □

Based on Theorem 6.3, T of free CNP set can be obtained as follows.

Theorem 6.4 *For any CNP graph $G = \{C, \mathcal{E}\}$ with the minimum degree $\Delta_{\min} = \min \{d (c) \,|\, c \in C\}$,*

$$T \leq \frac{2 \,|C|}{\Delta_{\min} + 2}.$$

This is a direct result from [15], where the proof can also be found. The result highly relies on the feature of claw-free CNP graphs. According to above theorem, it is seen that given the same number of CNPs, the chain topology has the largest T. Specially, for the chain topology, $k \geq 1$, if $|C| = 2k$, $T = |C|/2$; otherwise, $T = (|C| + 1)/2$.

Finally, we make the following two remarks.

- It is worth noting that although the proposed model and algorithm are illustrated in the simple scenario, where only pairs of NAP are partially overlapped, it is applicable for general scenarios. We extend the concept of CNP to incorporate overlapping areas formed by multiple networks. To illustrate this, we present the mappings from two different network topologies to the corresponding CNPs or players of the game. The CNP in Fig. 6.1, which we call it chain topology, illustrates roadside cellular network base stations. In Fig. 6.2, the nestification topology represents the cellular–WLAN integration scenario, the macro/pico/femto deployment or multimode small cells in cellular networks. The overlapping topology represents the conventional cellular network distribution, where networks partially overlapped with each other. As can be seen, there is an overlapping area covered by networks A, B, and D, and an overlapping area covered by networks A, B, and C in sub-figure a and b, respectively. The corresponding CNPs are extended to include the three networks. The relevant strategy are extended as $\mathbf{x}_{\langle A,B,D \rangle}$ and $\mathbf{x}_{\langle A,B,C \rangle}$, respectively. We argue that the algorithms and analysis except for Theorem 6.2 are still valid for the extended CNP cases. In the rest of this chapter, we use $\langle n_i, n_j \rangle$ to denote a CNP consisting of networks n_i and n_j, and use c to denote a CNP without specifying the involved networks.
- In order to find $\mathbf{x}'_{\langle n_i, n_j \rangle}$ in each iteration, the selected CNP $\langle n_i, n_j \rangle$ has to enumerate $\left| \mathbf{X}_{\langle n_i, n_j \rangle} \right|$ strategies on the proposed algorithms. If the user number is relatively large, a weak search where a user distribution better than current one rather than

the best one can be adopted. In this circumstance, the updates in both LIA and E-LIA are replaced by (6.10), where $\tilde{\mathbf{x}}_{\langle n_i,n_j\rangle}$ is current user distribution. To find a better user distribution, the enumeration number is in the range $\left[1, \left|\mathbf{X}_{\langle n_i,n_j\rangle}\right|\right]$, that is, the enumeration number is upper bounded by $\left|\mathbf{X}_{\langle n_i,n_j\rangle}\right|$. The search complexity is relieved in weak exhaust search at the cost of slower convergence speed. In the context of weak search, Theorems 6.1 and 6.2 still hold due to the finite improvement property of the finite ordinary potential game.

$$
\begin{aligned}
\mathbf{x}'_{\langle n_i,n_j\rangle} \in \Big\{ \mathbf{x}_{\langle n_i,n_j\rangle} \Big| & \, U_{\langle n_i,n_j\rangle}\left(\mathbf{x}_{\langle n_i,n_j\rangle}, \mathbf{x}_{-\langle n_i,n_j\rangle}\right) \\
& > U_{\langle n_i,n_j\rangle}\left(\tilde{\mathbf{x}}_{\langle n_i,n_j\rangle}, \mathbf{x}_{-\langle n_i,n_j\rangle}\right), \mathbf{x}_{\langle n_i,n_j\rangle} \in \mathbf{X}_{\langle n_i,n_j\rangle} \Big\}.
\end{aligned}
\tag{6.10}
$$

6.5　Simulation Results

In this section, some simulation results are presented.

6.5.1　Simulation Setting

The considered network consists of partially overlapped WLAN and LTE NAPs. The capacity of each LTE cell can be 10–20 Mbps, while the capacity of each WLAN (e.g., IEEE 802.11a) can be 6–54 Mbps [16]. Several users randomly locate in the networks. Each user falls into one of the three types, that is, brittle traffic user, stream traffic user, and elastic traffic user.

The user demand functions for brittle traffic users, stream traffic users, and elastic traffic users [8] are, respectively,

$$
f(\theta) = \begin{cases} 1, & \theta \geq \theta_t \\ 0, & \theta < \theta_t \end{cases},
\tag{6.11}
$$

$$
g(\theta) = 1 - \exp\left(-\frac{\lambda_2 \theta^2}{\lambda_1 + \theta}\right),
\tag{6.12}
$$

$$
h(\theta) = 1 - \exp\left(\frac{\delta\theta}{\theta_{\text{peak}}}\right),
\tag{6.13}
$$

where θ_t is the throughput threshold of the brittle traffic user, θ_{peak} is the peak throughput of the elastic traffic user, λ_1, λ_2, and δ are relevant parameters. In the simulation, $\theta_t = 1$ Mbps, $\lambda_1 = 30$, $\lambda_2 = 0.3$, $\delta = -0.6$, $\theta_{peak} = 2$ Mbps.

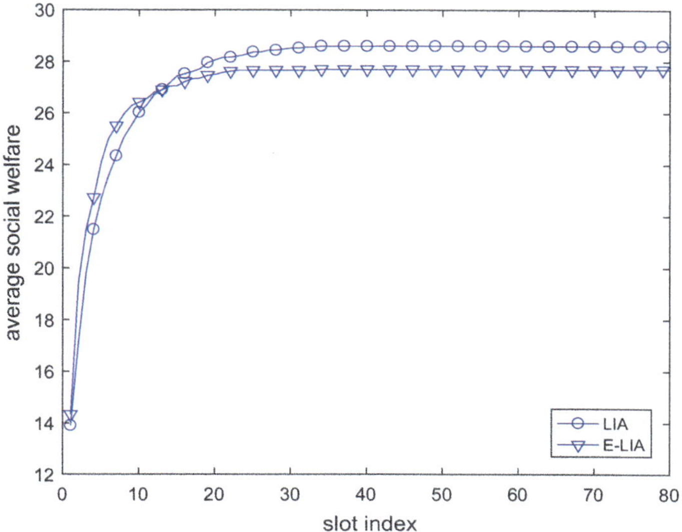

Fig. 6.4 Convergence behavior of the algorithms

6.5.2 Results

The convergence performance of the proposed algorithms is simulated in a chain topology composed of three LTE cells $lte1, lte2, lte3$, and three WLANs $wlan1$, $wlan2, wlan3$. The CNP set is

$$C = \{\langle lte1, wlan1 \rangle, \langle wlan1, lte2 \rangle, \langle lte2, wlan2 \rangle, \langle wlan2, lte3 \rangle, \langle lte3, wlan3 \rangle\}$$

and the regular user distribution are represented as $\mathbf{y}^{lte1}$, $\mathbf{y}^{lte2}$, $\mathbf{y}^{lte3}$, $\mathbf{y}^{wlan1}$, $\mathbf{y}^{wlan2}$, $\mathbf{y}^{wlan3}$. The maximal user number of each type for regular user or overlapping area user is set as 6. The weight is set as $w_b = 1, w_s = 1.5, w_e = 1$.

The social welfare averaged by 100 samples of LIA and E-LIA is shown in Fig. 6.4. We can observe that both algorithms could converge fast, i.e., less than 50 slots for the considered topology. Approximately, it costs 40 slots and 30 slots to converge for LIA and E-LIA, respectively. However, note that the finally converged social welfare of LIA is larger than that of E-LIA. To compare the convergence speed more clearly, the empirical CDF of convergence speed of the two algorithms are presented in Fig. 6.5. It is obvious that E-LIA converges faster than LIA. The two figures indicates that the simultaneous update of free CNPs does improve the convergence speed, which, however, achieves poor social welfare probably due to the reduced optimization space.

The performances of LIA and another three algorithms are simulated in a nestification topology similar to Fig. 6.2a, including a LTE cell $lte1$ and three WLANs

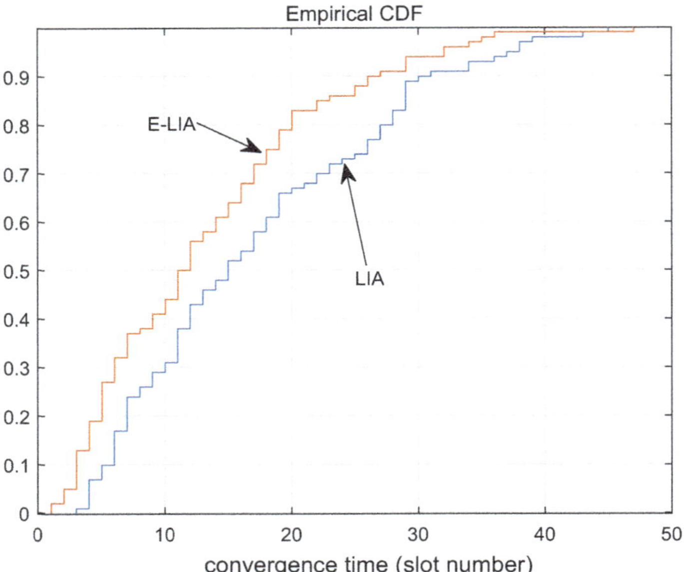

Fig. 6.5 Empirical CDF of algorithm convergence time

$wlan1, wlan2, wlan3$. The CNP set is

$$C = \{\langle lte1, wlan1, wlan2\rangle, \langle lte1, wlan2\rangle, \langle lte1, wlan3\rangle\}$$

and the regular user distribution is only $\mathbf{y}^{lte1}$. The considered algorithms are: (1) Individual reinforcement learning algorithm. Each individual user maintains and adjusts a Q-value for each available network reflecting the expected payoff for selecting the corresponding network. The Q-value is updated by the user's interaction with the environment. Users select the network with the largest Q-value while keeping an exploration step with a small probability in each iteration [11]. (2) Demand contention algorithm. Each user m selects the network trying to maximize the user utility $u_m(\theta_m, c_m)$. In each iteration, only one randomly selected user is allowed to select the network maximizing its utility, which is known as the best response. The demand contention algorithm can also converge to multiple stable points according to our simulation. (3) Social best response algorithm. Similar to the demand contention algorithm, the only difference is user utility. In each iteration, only one randomly selected user is allowed to select the network maximizing the social welfare of the system.

Figure 6.6 shows the performance comparison averaged by 200 samples under five different resource schedule weights: [0.5, 0.8, 1], [0.8, 0.5, 1], [1, 1, 1], [1.2, 2, 1], and [2, 1.2, 1]. All algorithms have a similar variation trend as the resource schedule policy changes. It seems that the maximal and minimal average social welfare is

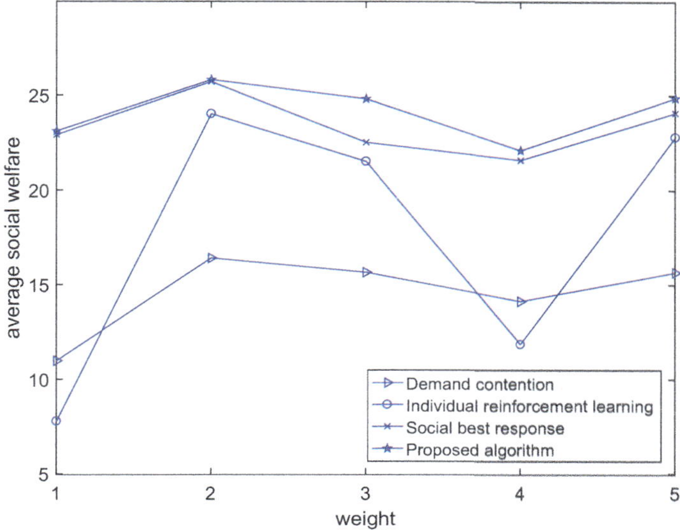

Fig. 6.6 Algorithm comparison with different resource schedule weights

achieved at $\mathbf{W}_2$ and $\mathbf{W}_4$, respectively. This phenomenon may put forward the joint network selection and user weight optimization problem, which could be one of our future works. Due to user competition, the demand contention and individual reinforcement learning algorithms are significantly worse than the proposed algorithm, especially for $\mathbf{W}_1$ and $\mathbf{W}_4$. Although the social welfare maximization is the goal of users and the global information is required in the social best response, it achieves poor performance than the proposed algorithms in all cases. We infer that the number of decision makers (user terminals) in the social best response is much larger than that in the proposed algorithms, which indicates that the dimension of the optimization variables in the former case is much larger. Therefore, the probability of achieving the global optimum is relatively smaller in the social best response. The result illustrates that the proposed algorithm can provide promising performance regardless of the specific weight vectors in the resource schedule policy.

We further check the effect of network loads on the performance of all algorithms. The load here denotes the total user number in all the networks. We increase the user number to get four different load levels for each topology, where the load increases from level 1 to level 4. Specifically, the user numbers are roughly in the range of 30~60. Figure 6.7 revealed that except for individual reinforcement learning, the social welfare of the algorithms increase as the load increases, although the growth speeds are different. Obviously, the proposed algorithm outperforms the other algorithms. The social welfare of the reinforcement learning algorithm degrades significantly when the network load becomes heavy, e.g., beyond level 3. The social best response algorithm also achieves less social welfare than the proposed algorithm in all cases.

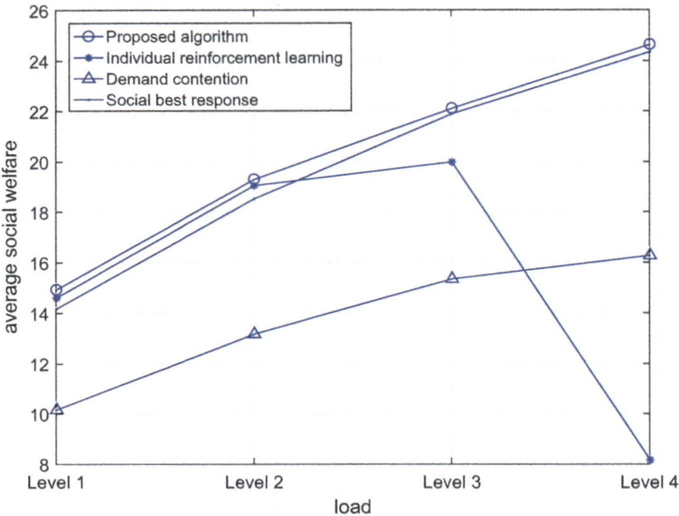

Fig. 6.7 Algorithm comparison with different user loads

6.6 Conclusion

This chapter studied the network selection for multiple users with heterogeneous demand. To model the problem, a novel localized self-organization game is proposed, where CNPs are abstracted as players. Based on the game, two simple algorithms LIA and E-LIA are designed. The convergence and performance of the proposed algorithms are analyzed. It is verified that the proposed algorithms could achieve fair performance with limited optimization cost.

References

1. Xue P, Gong P, Park J et al (2012) Radio resource management with proportional rate constraint in the heterogeneous networks. IEEE Trans Wirel Commun 11(3):1066–1075
2. Prasad N, Zhang H, Zhu H et al (2014) Multiuser scheduling in the 3GPP LTE cellular uplink. IEEE Trans Mob Comput 13(1):130–145
3. Keshavarz-Haddad A, Aryafar E, Wang M, Chiang M (2017) HetNets selection by clients: convergence, efficiency, and practicality. IEEE ACM Trans Netw 25(1):406–419
4. Malanchini I, Cesana M, Gatti N (2013) Network selection and resource allocation games for wireless access networks. IEEE Trans Mob Comput 12(12):2427–2440
5. Feng X, Gan X et al (2017) Distributed cell selection in heterogeneous wireless networks. Comput Commun 109:13–23
6. Nguyen DD, Nguyen HX, White LB (2017) Reinforcement learning with network-assisted feedback for heterogeneous RAT selection. IEEE Trans Wirel Commun 16(9):6062–6076
7. Du Z, Wu Q, Yang P, Yuhua Xu, Yao YD (2014) User-demand-aware wireless network selection: a localized cooperation approach. IEEE Trans Veh Technol 63(9):4492–4507

8. Rakocevic V, Griffiths J, Cope G (2001) Performance analysis of bandwidth allocation schemes in multiservice IP networks using utility functions. In: Proceedings of the 17th international teletraffic congress (ITC)
9. Deb S, Nagaraj K, Srinivasan V (2011) MOTA: engineering an operator agnostic mobile service. MobiCom 2011
10. Arnborg S (1985) Efficient algorithms for combinatorial problems on graphs with bounded decomposability-a survey. BIT Numer 25(1):1–23
11. Niyato D, Hossain E (2009) Dynamics of network selection in heterogeneous wireless networks: an evolutionary game approach. IEEE T Veh Technol 58(4):2008–2017
12. Costa-Pérez X et al (2013) Latest trends in telecommunication standards. ACM Comput Commun Rev
13. Monderer D, Sharpley LS (1996) Potential games. Games Econ Behav 14:124–143
14. Diestel R (2000) Graph theory, 2nd edn. Springer, New York, p 2000
15. Faudree RJ, Gould RJ, Jacobson MS et al (1992) On independent generalized degrees and independence numbers in K(l, m)-free graphs. Discrete Math 103:17–24
16. Raychaudhuri D, Mandayam NB (2012) Frontiers of wireless and mobile communications. Proc IEEE 100(4):824–840

Chapter 7
Exploiting User Demand Diversity: QoE Game and MARL Based Network Selection

Abstract This chapter studies distributed network selection for multiple user cases with heterogeneous demand. The key challenge is low system efficiency due to user competition. Motivated by the fact that the ultimate goal of communications is to serve users with personalized demand, we introduce a new concept of user demand diversity gain. This gain derives from the elaborate matching between user demand and radio resource, which cannot be directly attained in the existing throughput-centric optimization due to users' blindness in maximizing throughput. Aiming at obtaining this gain, we propose the user demand centric optimization, where users seek to maximize QoE. To model this problem, we propose a novel game formulation, QoE game. The properties of QoE game and QoE equilibrium and the definition of user demand diversity gain are presented. Two distributed QoE equilibrium learning algorithms, stochastic learning automata (SLA) based algorithm and trail and error (TE) based algorithm, are designed to achieve QoE equilibrium. Simulation results validate the existence of user demand diversity gain and the effectiveness of the proposed learning algorithms in improving the system efficiency and QoE fairness.

7.1 Introduction

Similar to the previous chapter, this chapter studies network selection for multiple user cases with heterogeneous demand. However, we attempt to design distributed algorithms. Compared to the centralized or hybrid approach in the previous chapter, the advantage of distributed solutions is low cost, scalability, and robustness, while the challenges are algorithm design. On one hand, due to resource sharing nature among users in one NAP, users' decision-making is coupled. Thus, it is crucial to ensure distributed algorithm to converge fast in such a nonstationary environment. On the other hand, beyond convergence, how to achieve high system efficiency under the distributed decision-making framework is another challenge.

Game theory is a powerful tool in analyzing distributed decision-making. Game-based network selection has received much attention in recent years [1, 2]. Widely applied game models include congestion game [3], evolution game [4], hierarchical game [5], etc.. In these works, various theories and mechanisms have been proposed

© Springer Nature Singapore Pte Ltd. 2020
Z. Du et al., *Towards User-Centric Intelligent Network Selection in 5G Heterogeneous Wireless Networks*,
https://doi.org/10.1007/978-981-15-1120-2_7

to guarantee the convergence, however, the efficiency issue is not fully analyzed. The challenge originates from the nature of distribution decision-making: the behaviors of autonomous users are driven by selfish utility function, which is not in accordance with the system global utility. Therefore, *"the tragedy of commons"* is a common outcome of distributed solutions.

To achieve fair system performance for a distributed solution, this chapter introduces QoE into game model and replace traditional QoS-centric optimization scheme with user demand centric optimization scheme to alleviate users' blind competition. On this basis, we exploit the subjectiveness characteristic of user demand to further improve system efficiency. Motivated by this idea, we model the network selection as a novel QoE game, analyze its properties, and prove that QoE equilibrium is more efficient than QoS equilibrium. Moreover, two game-based multi-agent RL (MARL) algorithms are proposed and evaluated. The main results of this chapter were presented in [6].

7.2 Motivation

7.2.1 *User Demand Diversity*

Most existing works in resource management and network selection maximize the throughput or QoS. However, the ultimate goal of wireless communications is to serve people with diverse demands. There could be a gap between maximizing QoS and QoE, especially considering the heterogeneous demand. The heterogeneity of user demand generally comes from two aspects. The first is the different performance requirements of diverse types of traffic or applications. As we have mentioned in previous chapters, video traffic, audio traffic, and file transfer have different requirements on performance parameters such as throughput, delay, and so on. Even for the same traffic type, the requirements can be different. For example, the following Table 7.1 shows that the bandwidth requirement of video call varies significantly with the video quality. The second aspect is users' preference, that is, due to the subjective feature, users' perception could be different even with the same QoS level.

In wireless networks, with the limited radio resource, if users do not selfishly and blindly maximize QoS but just satisfy their demand will alleviate the competition and improve system efficiency. We call the "potential system efficiency improvement derived from an optimal match between diverse user demand and network resource" *user demand diversity*. In the following, we study how to achieve it in distributed learning context.

Table 7.1 Bandwidth requirements for Skype

Call type	Minimum download/upload speed	Recommended download/upload speed
Voice call	30 kbps/30 kbps	100 kbps/100 kbps
Video call/screen sharing	128 kbps/128 kbps	300 kbps/300 kbps
Video call (high quality)	400 kbps/400 kbps	500 kbps/500 kbps
Video call (HD)	1.2 Mbps/1.2 Mbps	1.5 Mbps/1.5 Mbps
Multiplayer video (3 people)	512 kbps/128 kbps	2 Mbps/512 kbps
Multiplayer video (>7 people)	4 Mbps/128 kbps	8 Mbps/512 kbps

7.2.2 How to Exploit User Demand Diversity?

While existing methods neglected the diverse user demand and its effect on resource management, we propose user demand centric optimization method. Its idea is that user maximizes QoE level rather than QoS or throughput quantity, which is motivated by the following two points:

- When a throughput or QoS improvement, e.g., 0.01 Mbps, is not large enough to improve a user's QoE level, it makes no difference to the user.
- Moreover, when a user's QoE cannot be improved anymore, a larger throughput would be meaningless.

The above subjective demand properties derive from the fact that a user can only feel limited levels of QoE rather than distinguish precisely quantitative throughput. It is worth noting that when users seek for maximizing QoE, the above properties can significantly relieve the blindness and aggression of users. That is, users compete for a larger throughput only when the QoE can be improved and a user will stop competing for a larger throughput once it already has the best QoE. In the rest of this chapter, we will attempt to analyze the properties of user demand centric optimization and how to achieve the user demand diversity gain in distributed network selection with autonomic and selfish users.

7.3 System Model and Problem Formulation

On the basis of system model, this section models the user demand centric network selection. Specifically, discrete QoE model is introduced and the network selection is formulated as a system-level QoE optimization problem.

7.3.1 System Model

Consider a heterogeneous wireless network with N NAPs coexisting in a given area. For simplicity, the involved NAPs can be generally classified as macrocell or small cell, where macrocells refer to common 3G, 4G, and 5G cellular network cells and small cells could be picocells and femtocells in cellular networks, APs in WLANs. The network set is denoted by $\mathcal{N} = \{1, ..., N\}$. A set of users $\mathcal{M} = \{1, ..., M\}$ locate in the networks. Each user has a set of available NAPs $\mathcal{A}_m \subseteq \mathcal{N}$. In particular, users located in overlapping areas of networks can access any one of the networks, but access only one network at any given time.

Specifically, a user's throughput is determined by the physical layer data rate, the load on the associated network, and the network-side resource allocation policy. For network-side resource allocation policy, we assume that the proportional fairness and soft-QoS-based service differentiation are adopted in networks. These schemes are widely used in cellular networks including LTE-A. In this context, the average throughput θ_m of user $m \in \mathcal{M}$ given the accessed network n is derived by

$$\theta_m = \frac{w_m R_{m,n}}{W_n}, \tag{7.1}$$

where $R_{m,n}$ is the physical layer data rate between user m and network n, w_m is user m's weight, $W_n = \sum_{i \in \mathcal{M}_n} w_i$ is the total user weight of network n indicating the load of the network, $\mathcal{M}_n$ is the set of users associated with network n. The above throughput model incorporates several practical considerations. First, the data rate $R_{m,n}$ reflects the physical layer characteristics such as radio channel condition, modulation, and coding scheme. Second, the discount factor $\frac{w_m}{W_n}$ on $R_{m,n}$ shows the nature of multiple user resource sharing. Third, the user weight here differentiates users in the same network with diverse traffic or applications, which can be seen as an indication of application layer information. The model is similar to the one in [7] and we generalize the user weight from 1 to w_m.

7.3.2 Discrete QoE Model

As we have analyzed, maximizing quantitative QoS will lead to blind competition. Following the idea of user demand centric optimization, we believe the key to alleviate competition is using discrete QoE. Denote the selected network of user m by $a_m \in \mathcal{A}_m$, the network selection strategy profile of all users by $\mathbf{a} = (a_1, ..., a_M)$ with the joint strategy profile set $\mathcal{A} = \underset{m \in \mathcal{M}}{\times} \mathcal{A}_m$. We define a QoE function $q(\cdot)$ to represent the user demand, which maps the strategy profile $\mathbf{a}$ to QoE levels. In particular, considering the subjectiveness of user perception, we can adopt the following two QoE models.

(1) **Binary satisfaction based QoE model**: In binary satisfaction based model, user's QoE can be classified into two levels including "satisfactory" and "unsatisfactory". This binary satisfaction model characterizes the hard QoS requirement of users in specific traffic or services such as the target SINR requirement in power control [8], the bandwidth requirement for brittle traffic [9]. Normally, there is a threshold on the QoS requirement, beyond which the user satisfaction is well enough, otherwise, it is unacceptable. For this, the QoE function can be as follows:

$$q_m\left(\mathbf{a}\right) = \begin{cases} \text{satisfactory}, & \mathbf{a} \in Q_{m,1} \\ \text{unsatisfactory}, & \mathbf{a} \in Q_{m,2}, \end{cases} \tag{7.2}$$

where $Q_{m,1}$ and $Q_{m,2}$ are the sets of strategy profiles corresponding to "satisfactory" and "unsatisfactory" levels, respectively, $Q_{m,1} \bigcap Q_{m,2} = \emptyset$, $Q_{m,1} \bigcup Q_{m,2} = \mathcal{A}$. For clarity, we denote $Q_m = \left(Q_{m,1}, Q_{m,2}\right)$. Note that this model is similar with the model in satisfaction equilibrium [8].

(2) **MOS-based QoE model**: One limitation of the binary satisfaction based QoE model is that it seems too hard: either satisfactory or unsatisfactory without intermedial states. In many cases, it calls for finer representations for user perceived experience. Actually, both in the academia and the industry community, the most widely used metric to capture user satisfaction is mean opinion score (MOS), as shown in Table 7.2. The MOS has five levels representing users' subjective estimation "excellent", "good", "fair", "poor", and "bad", where the MOS is related with the impairment of related QoS. Although MOS is originally used for voice quality estimation, it had been extended to estimate the quality of video, file transfer, etc., in [10]. Inspired by the MOS, we propose the MOS-based QoE model, where the QoE function can be as follows:

$$q_m\left(\mathbf{a}\right) = \begin{cases} \text{Excellent} & \mathbf{a} \in Q_{m,1} \\ \text{Good} & \mathbf{a} \in Q_{m,2} \\ \text{Fair} & \mathbf{a} \in Q_{m,3} \\ \text{Poor} & \mathbf{a} \in Q_{m,4} \\ \text{Bad} & \mathbf{a} \in Q_{m,5}, \end{cases} \tag{7.3}$$

where $Q_{m,i}, i = 1, ..., 5$ are the sets of strategy profiles corresponding to different user QoE levels, $\bigcap_{i=1}^{5} Q_{m,i} = \emptyset$, $\bigcup_{i=1}^{5} Q_{m,i} = \mathcal{A}$. We also denote $Q_m = \left(Q_{m,1}, Q_{m,2}, Q_{m,3}, Q_{m,4}, Q_{m,5}\right)$.

The above QoE levels can be determined as follows. Resorting to the well-known throughput to MOS mapping in [11], we have

$$MOS = \lambda \log_{10}\left(\gamma\theta\right), \tag{7.4}$$

where θ is user's throughput, λ, γ are parameters dependent on specific maximal and minimal throughput demand of the user. We derive $Q_{m,i}$ in terms of the achieved throughput under $\mathbf{a}$, $\theta_m\left(\mathbf{a}\right)$, by dividing the MOS into five partitions as follows:

Table 7.2 Mean opinion score (MOS)

MOS	Quality	Impairment
1	Bad	Very annoying
2	Poor	Annoying
3	Fair	Slightly annoying
4	Good	Perceptible but not annoying
5	Excellent	Imperceptible

$$Q_{m,5} = \left\{ \mathbf{a} \,\middle|\, \lambda_m \log_{10}\left[\gamma_m \theta_m(\mathbf{a})\right] \geq 4.5 \right\}$$
$$Q_{m,4} = \left\{ \mathbf{a} \,\middle|\, 3.5 \leq \lambda_m \log_{10}\left[\gamma_m \theta_m(\mathbf{a})\right] < 4.5 \right\}$$
$$Q_{m,3} = \left\{ \mathbf{a} \,\middle|\, 2.5 \leq \lambda_m \log_{10}\left[\gamma_m \theta_m(\mathbf{a})\right] < 3.5 \right\} \tag{7.5}$$
$$Q_{m,2} = \left\{ \mathbf{a} \,\middle|\, 1.5 \leq \lambda_m \log_{10}\left[\gamma_m \theta_m(\mathbf{a})\right] < 2.5 \right\}$$
$$Q_{m,1} = \left\{ \mathbf{a} \,\middle|\, \lambda_m \log_{10}\left[\gamma_m \theta_m(\mathbf{a})\right] < 1.5 \right\}$$

which indicates that the QoE level can be determined by four throughput thresholds. Note that such QoE level mapping is especially suitable for video or audio applications involving users' subjective evaluations, while it may incur deviations in other applications such as file transfer.

Some other important factors such as delay or a combination of multiple factors can also be mapped to QoE level. Note that we mainly consider the throughput-QoE mapping since the throughput is the determinant factor in QoE and is highly related to RRM. To differentiate diverse user demand, the QoE function $q_m(\mathbf{a})$ is user specific, i.e., λ_m and γ_m are set according to user demand.

7.3.3 System-Level QoE Optimization

To the best of our knowledge, current QoE evaluation models and approaches are limited to single user case. There is a lack of system performance evaluation from QoE perspective. The social welfare in terms of the sum utility of users is not suitable due to the discrete QoE level. To fill this void, we extend the QoE to multiple user cases and propose the system-level QoE on the basis of the above user QoE model.

Definition 7.1 Given a strategy profile $\mathbf{a}$, the system performance is measured by system-level QoE, defined as $\mathbf{D}(\mathbf{a}) = (d_5(\mathbf{a}), d_4(\mathbf{a}), d_3(\mathbf{a}), d_2(\mathbf{a}), d_1(\mathbf{a}))$, where $d_k(\mathbf{a}) = \sum_{m \in M} I\{\kappa_m(\mathbf{a}) \geq k\}$, $k = 1, ..., 5$ denotes the number of users with a QoE level no worse than $Q_{m,k}$, $I\{\cdot\}$ is the indicator function, $\kappa_m(\mathbf{a})$ is the QoE level index of user m, i.e., $\mathbf{a} \in Q_{m,\kappa_m(\mathbf{a})}$.

Here, $d_1(\mathbf{a}) = M$, $\forall \mathbf{a}$. The above definition possesses several advantages. First, it implicitly indicates the user numbers at different QoE levels. Note that the number of users with level 5, "Excellent", is $d_5(\mathbf{a})$, the number of users with level $1 \leq k < 5$ is $d_{k+1}(\mathbf{a}) - d_k(\mathbf{a})$. Second, this metric provides flexible performance evaluations. From the system point of view, one may not care the exact number of users at each QoE level, rather, the number of users with QoE better than some threshold such as "Good" or "Fair" can be practical and important. In this situation, $d_4(\mathbf{a})$ and $d_3(\mathbf{a})$ are just the desirable metrics. Third, the gap between any two strategy profiles provides direct performance comparison result. For instance, $\mathbf{D}(\mathbf{a}) - \mathbf{D}(\mathbf{a}') = (0, 1, -1, 0, 0)$. If the number of users with QoE level equal or better than "Good" is the efficiency measure, $\mathbf{a}$ is better than $\mathbf{a}'$; while $\mathbf{a}'$ is better than $\mathbf{a}$ if the number of users with QoE level equal or better than "Fair" is the efficiency measurement.

We specify the used performance comparison criterion, which is based on the lexicographical value of accumulative user distributions. Specifically, $\mathbf{a}$ is better (more efficient) than $\mathbf{a}'$ if the lexicographical value of $\mathbf{D}(\mathbf{a})$ is higher than that of $\mathbf{D}(\mathbf{a}')$, that is, there exists $1 < j \leq 5$ satisfying $d_i(\mathbf{a}) = d_i(\mathbf{a}')$ for $j < i \leq 5$ and $d_j(\mathbf{a}) > d_j(\mathbf{a}')$. $\mathbf{a}$ is strictly better than $\mathbf{a}'$ if $d_i(\mathbf{a}) \geq d_i(\mathbf{a}')$, $1 \leq i \leq 5$ and there exists at least one $1 < j \leq 5$ satisfying $d_j(\mathbf{a}) > d_j(\mathbf{a}')$. $\mathbf{a}$ is equivalent to $\mathbf{a}'$ if $\mathbf{D}(\mathbf{a}) = \mathbf{D}(\mathbf{a}')$. The above three types of relationships are denoted as $\mathbf{a} \rhd \mathbf{a}'$, $\mathbf{a} \rhd \rhd \mathbf{a}'$, and $\mathbf{a} \doteq \mathbf{a}$, respectively. The optimal (most efficient) strategy profile $\mathbf{a}^*$ satisfies

$$\mathbf{a}^* \rhd \mathbf{a} \quad \text{or} \quad \mathbf{a}^* \doteq \mathbf{a}, \forall \mathbf{a} \in \mathcal{A}. \tag{7.6}$$

Therefore, the optimization goal is to find the optimal strategy $\mathbf{a}^*$.

For example, suppose that there are 10 users in a heterogeneous wireless network and three strategy profiles $\mathbf{a}_1$, $\mathbf{a}_2$, and $\mathbf{a}_3$ with $\mathbf{D}(\mathbf{a}_1) = (3, 5, 6, 8, 10)$, $\mathbf{D}(\mathbf{a}_2) = (3, 5, 5, 8, 10)$ and $\mathbf{D}(\mathbf{a}_3) = (3, 4, 7, 9, 10)$. Their relationships are $\mathbf{a}_1 \rhd \mathbf{a}_3$, $\mathbf{a}_1 \rhd \rhd \mathbf{a}_2$, and $\mathbf{a}_2 \rhd \mathbf{a}_3$. Note that the intuition behind the above criterion is that the higher QoE level has a dominant effect in efficiency. Other criteria may also be adopted in the performance comparison.

7.4   QoE Game-Based Network Selection

In this section, a novel QoE game model is proposed to analyze the formulated problem. Then, general and specific properties in network selection are studied. Finally, the user demand diversity gain is defined based on QoE equilibrium.

7.4.1   QoE Game Model

For comparison, following the existing QoS-centric optimization idea, we first give a so-called QoS game or throughput game $G_\theta = \left(\mathcal{M}, \{\mathcal{A}_m\}_{m \in \mathcal{M}}, \{\theta_m\}_{m \in \mathcal{M}}\right)$. Then, on the basis of the above discrete QoE model, we formally define QoE game G_{QoE}

for general resource management (not limited to current network selection problem) as follows.

Definition 7.2 The QoE game of the distributed resource management is $G_{QoE} = \left(\mathcal{M}, \{\mathcal{A}_m\}_{m \in \mathcal{M}}, \{Q_m\}_{m \in \mathcal{M}} \right)$.

We highlight the following properties of the QoE game:

- Subjective insensitivity to fine distinction of QoS. Note that each $Q_{m,i} \neq \emptyset$ may correspond to a subset of $\mathcal{A}$, where different strategies may lead to different QoS but the same QoE level for the user. Actually, users' subjective experience can "tolerate" the QoS variation in a limited range. For instance, although 1.01 Mbps is larger than 1 Mbps in the throughput, they may result in the same QoE level to the user. In a word, the insensitivity to QoS variation originates from the subjective QoE evaluation.
- Bounded greed. Once $\mathbf{a}$ belongs to $Q_{m,1}$, user m's QoE level cannot be improved any more and the user will stop pursuing better QoS. This property results from the fact that when the experienced QoS is good enough, users' subjective experience is indifferent on the specific QoS.
- Diverse user demand. Q_m is distinct for each user. Therefore, even given the same QoS, two users can have different QoE levels. This is mainly due to the diversity in running traffic types, user preferences, etc.. For instance, 0.1 Mbps throughput is good enough for a voice traffic user, while it may be annoying for a video traffic user.

Clearly, users seek for the possible highest QoE level in the QoE game. In the throughput game, the preference order is built on the quantifiable user utility. The QoE game differs from most existing games in that the QoE level Q_m is an ordinal qualitative scale by nature, not a quantitative scale. This restriction indicates that utility-based analysis cannot be directly used here. Therefore, we define the preference order according to QoE level. Given any two strategy profiles $\mathbf{a}$ and $\mathbf{a}'$, $\mathbf{a} \in Q_{m,i}$, $\mathbf{a}' \in Q_{m,j}$, user m weakly prefers $\mathbf{a}$ to $\mathbf{a}'$ if $i \geq j$. This binary relationship is denoted by $\mathbf{a} \succeq \mathbf{a}'$. Let $\mathbf{a}_{-m}$ denotes the strategy set of all users except for m, then $\mathbf{a}$ can be denoted by $(a_m, \mathbf{a}_{-m})$.

In the following, we first define the pure strategy Nash equilibrium for comparison. Then, we define the pure strategy QoE equilibrium and perfect pure strategy QoE equilibrium.

Definition 7.3 A strategy profile $\mathbf{a}$ is a pure strategy Nash equilibrium (PNE) of G_θ if no user can improve its throughput by unilaterally deviating its strategy, i.e.,

$$\theta_m\left(a_m, \mathbf{a}_{-m}\right) \geq \theta_m\left(a'_m, \mathbf{a}_{-m}\right),$$

$\forall m \in \mathcal{M}, \forall a'_m \in \mathcal{A}_m$, where $\theta_m(\mathbf{a})$ is user m's throughput given strategy profile $\mathbf{a}$.

Definition 7.4 A strategy profile $\mathbf{a}$ is a pure strategy QoE equilibrium (PQE) of G_{QoE} if no user can improve its QoE level by the throughput increase derived from unilaterally deviating its strategy, i.e.,

$$(a_m, \mathbf{a}_{-m}) \succeq \left(a'_m, \mathbf{a}_{-m}\right),$$

$\forall m \in \mathcal{M}, \forall a'_m \in \mathcal{A}_m$.

Definition 7.5 A strategy profile $\mathbf{a}$ is a perfect pure strategy QoE equilibrium (PPQE) of G_{QoE} if all users experience the "Excellent" QoE level, simultaneously, i.e., $\hat{\mathbf{a}} \in Q_{m,5}, \forall m \in \mathcal{M}$.

We can see that the definition of PQE is similar to the existing pure strategy Nash equilibrium (PNE). In addition, the PPQE is a special case of PQE, where the heterogeneous wireless network resource is sufficient to accommodate all users. The PPQE is similar to the satisfaction equilibrium in [8].

In systems where the PQE does not exist, we define another relaxed equilibrium-mixed QoE equilibrium. Denote the mixed strategy profile $\pi = (\pi_1, ..., \pi_\mathbf{M}) \in \Pi$, where $\pi_m = \left(\pi_{m,1}, \pi_{m,2}, ..., \pi_{m,|A_m|}\right) \in \Pi_m$ is the probability distribution of user m's decision on his strategy set, $\pi_{m,k} \geq 0$ with $k = 1, 2, ..., 5, \sum_{k=1}^{5} \pi_{m,k} = 1$ and $k = 1, 2, \sum_{k=1}^{2} \pi_{m,k} = 1$ are the probabilities of user m selecting strategy k in binary satisfaction based and MOS-based QoE models, respectively. Given the strategy profile π, the probability vector of user m being at each QoE level in the above two QoE models are denoted as $\mathbf{P}_m(\pi) = \left(p_{m,1}(\pi), p_{m,2}(\pi), ..., p_{m,5}(\pi)\right)$ and $\mathbf{P}_m(\pi) = \left(p_{m,1}(\pi), p_{m,2}(\pi)\right)$, respectively, where $p_{m,k}$ is the probability of achieving QoE level k for user m.

Definition 7.6 Denote the QoE weight vector of user m in the two QoE models as $\mathbf{X}_m = \left(x_{m,1}, x_{m,2}, ..., x_{m,5}\right)'$ and $\mathbf{X}_m = \left(x_{m,1}, x_{m,2}\right)'$, respectively. Define the expected QoE result of user m as $\mathbf{P}_m \mathbf{X}_m$. A strategy profile $\hat{\pi}$ is a mixed strategy QoE equilibrium (MQE) if and only no user can improve its expected QoE result by unilaterally deviating his mixed strategy, i.e.,

$$\mathbf{P}_m\left(\hat{\pi}\right) \mathbf{X}_m \geq \mathbf{P}_m\left(\pi'_m, \hat{\pi}_{-m}\right) \mathbf{X}_m$$

where Π_m is the set of all possible mixed strategy of user m, $\Pi = \underset{m \in \mathcal{M}}{\times} \Pi_m, \pi_{-m}$ is the mixed strategies of all users except for m, π is also denoted as (π_m, π_{-m}).

Different from conventional mixed strategy Nash equilibrium which is evaluated by the average utility, the MQE is evaluated by the probability. This is because that the QoE levels have no explicit utilities. It is easy to find that the PQE is a special case of MQE, where the mixed strategy of $\forall m \in \mathcal{M}$ is a unitary vector with $\hat{\pi}_{m,k_m} = 1$. The QoE weight vector reflects users' assessment on the QoE levels, which is key to the expected QoE result. In particular, we can envision two types of QoE weight vectors leading to different explanations of MQEs.

Type 1: There exists a target QoE level k_m with (1) $k_m = 1$ for binary satisfaction based QoE model and (2) $1 \leq k_m < 5$ for MOS-based QoE model, satisfying that $x_{m,k} = 1$, if $k \leq k_m$, otherwise, $x_{m,k} = 0$.

Type 2: The QoE weight vector satisfying $x_{m,k} \gg x_{m,k+1}, k = 1, 2, ..., 4$ for MOS-based QoE model.

Note that in the first type, each k_m is specified by the individual user. Additionally, users may change the QoE expectation by adjusting the target QoE levels. We exclude the case $k_m = 2$ and $k_m = 5$ for binary satisfaction based QoE model and MOS-based QoE model, respectively, since these settings indicate that $P\{\mathbf{a} \in Q_{m,k}, k \leq k_m \,|\, \pi\} = 1$, $\forall \pi \in \Pi$, that is, the user m is indifferent to users' strategies. The MQE in the first type essentially indicates that users maximize the probability of achieving no worse QoE than the target level. Actually, since users are not transmitting all the time, the QoE maximization in probability form is acceptable. On the other hand, the second type of QoE weight vector actually indicates that the MQE in this context is derived by comparing the lexicographical value of mixed strategies. It implies that a higher QoE level is overwhelmingly superior.

In addition to system efficiency, fairness is another important metric in evaluating the equilibrium. Due to the diversity in user demand, the fair throughput distribution among users may not lead to fair QoE and vice versa. To this end, a new fairness metric J_{QoE} based on the Jain' s fairness index (JFI) is derived as

$$
J_{QoE}(\mathbf{a}) = \frac{\left(\sum\limits_{m \in \mathcal{M}} \kappa_m(\mathbf{a}) \right)^2}{M \sum\limits_{m \in \mathcal{M}} [\kappa_m(\mathbf{a})]^2},
\tag{7.7}
$$

where $\kappa_m(\mathbf{a})$ is the QoE level index of user m under strategy $\mathbf{a}$. Note that when all users have the same QoE index level, $J_{QoE}(\mathbf{a})$ arrives at the largest value 1. This indicates that when $J_{QoE}(\mathbf{a})$ approaches 1, the corresponding QoE fairness improves.

7.4.2 Game Analysis

This subsection summarizes some *general* properties of QoE game for resource management. Generally, there is no result on the existence of PQE, which mainly depends on the specific system.

Proposition 7.1 *The QoE game G_{QoE} has no general result on the existence of PQE depending on the specific problem and system setting.*

Remark 7.1 We use an example to illustratively explain Proposition 7.1. We assume that there are three pairs of transmitter–receiver $\{A, B, C\}$ sharing 3 channels $\{ch_1, ch_2, ch_3\}$. They are close to each other that any two selecting the same channel are interfered with each other. Considering the diverse propagation loss, fading, shadow effect, landform, etc.., the interference link is asymmetric. That is, if the interference link gain between transmitter i and receiver j is denoted as $h_{i \to j}$, then $h_{i \to j} \neq h_{j \to i}$ for $i \neq j$, as shown in Fig. 7.1. Denote the resulting transmission rates of receivers in i and j as $Th_{j \to i}$ and $Th_{i \to j}$ when i and j use the same channel. Suppose the transmission rates follow the relationships: (1)

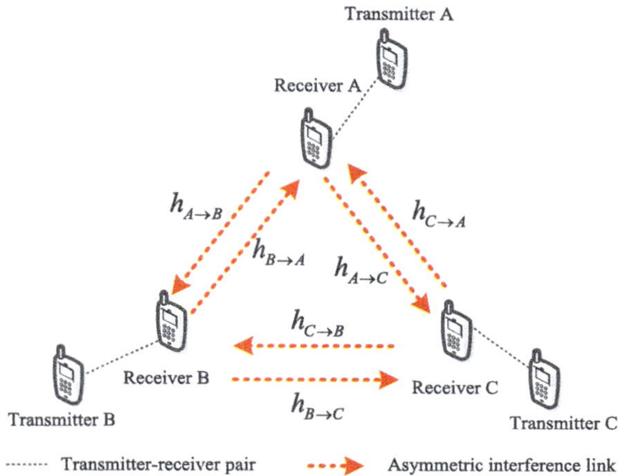

Fig. 7.1 An example to illustrate the case of no PQE exists. Considering the asymmetric interference, suppose $Th_{A\to B} = Th_{B\to C} = Th_{C\to A} = \sigma$, $Th_{A\to C} = Th_{C\to B} = Th_{B\to A} = 2\sigma$. If σ and 2σ lead to different QoE levels of all users, there is always one user that has the motivation to change the selected channel

$Th_{A\to B} = Th_{B\to C} = Th_{C\to A} = \sigma$, (2) $Th_{A\to C} = Th_{C\to B} = Th_{B\to A} = 2\sigma$, σ is a constant, (3) If all three users use the same channel, all users' transmission rates are smaller than σ. Obviously, if σ and 2σ lead to the same QoE level of all users, except the state that all three users select the same channel, all the other states are pure QoE equilibriums. On the contrary, if σ and 2σ lead to different QoE levels, there is no pure QoE equilibrium, since there is always one user in any two users selecting the same channel has the motivation to change channel to improve its QoE level.

The following result reveals the relationship between PQE and PNE.

Theorem 7.1 *Denote the PQE set in G_{QoE} as $\mathcal{A}_{PQE}$, the PNE set in G_θ as $\mathcal{A}_{PNE}$, then*

$$\mathcal{A}_{PQE} = \mathcal{A}_{PNE} \cup \mathcal{A}_\epsilon$$

where $\mathcal{A}_\epsilon$ is a subset of ϵ-equilibrium of the throughput game satisfying $\forall \mathbf{a} \in \mathcal{A}_\epsilon$, if $\kappa_m(\mathbf{a}) \neq 5$, then $\forall a'_m \in \mathcal{A}_m$, $\theta_m(a'_m, \mathbf{a}_{-m}) - \theta_m(a_m, \mathbf{a}_{-m}) < \epsilon_m(\mathbf{a})$, $\epsilon_m(\mathbf{a}) = \min\limits_{\mathbf{a}' \in Q_{m,\kappa_m(\mathbf{a})+1}} \theta_m(\mathbf{a}') - \theta_m(\mathbf{a})$, where $\theta_m(\mathbf{a})$ denotes user m's throughput under strategy profile $\mathbf{a}$.

Proof According to the definition of PNE, if $\mathbf{a}^*$ is a PNE in G_θ, no user can get a larger throughput by unilaterally deviating its strategy. As a result, no user can improve its QoE level, too. Hence, $\mathbf{a}^* \in \mathcal{A}_{PQE}$. On the other hand, for a PQE $\mathbf{a}$ that is not a PNE, it must satisfy the condition that the throughput increase of any user m with $\kappa_m(\mathbf{a}) \neq 5$ is bounded, such that a higher QoE level $\kappa_m(\mathbf{a}) + 1$ cannot be achieved. Otherwise, it contradicts with the definition of PQE. That is, $\forall m, \forall a'_m \in \mathcal{A}_m$, $\theta_m(a'_m, \mathbf{a}_{-m}) - \theta_m(a_m, \mathbf{a}_{-m}) < \min\limits_{\mathbf{a}' \in Q_{m,\kappa_m(\mathbf{a})+1}} \theta_m(\mathbf{a}') - \theta_m(a_m, \mathbf{a}_{-m})$. Therefore, $\mathcal{A}_{PQE} = \mathcal{A}_{PNE} \cup \mathcal{A}_\epsilon$. $\qquad\square$

Based on Theorem 7.1, we can get the following corollary.

Corollary 7.1 *Denote* $\mathbf{a}^*_{PQE}$ *as the best PQE and* $\mathbf{a}^*_{PNE}$ *as the best PNE both in terms of accumulative user distribution, that is,*

$$\mathbf{a}^*_{PQE} \in \left\{ \mathbf{a} \,\middle|\, \mathbf{a} \triangleright \mathbf{a}' \ or \ \mathbf{a} \doteq \mathbf{a}', \forall \mathbf{a}' \in \mathcal{A}_{PQE}, \mathbf{a} \in \mathcal{A}_{PQE} \right\}$$

$$\mathbf{a}^*_{PNE} \in \left\{ \mathbf{a} \,\middle|\, \mathbf{a} \triangleright \mathbf{a}' \ or \ \mathbf{a} \doteq \mathbf{a}', \forall \mathbf{a}' \in \mathcal{A}_{PNE}, \mathbf{a} \in \mathcal{A}_{PNE} \right\}$$

then $\mathbf{a}^*_{PQE} \triangleright \mathbf{a}^*_{PNE}$ *or* $\mathbf{a}^*_{PQE} \doteq \mathbf{a}^*_{PNE}$.

Theorem 7.1 shows that the QoE game is more robust than the throughput game in the sense that it possesses more equilibriums. Corollary 7.1 further concludes that under the QoE game framework, it is possible to attain equilibriums better than PNE. Actually, starting from a PNE $\mathbf{a}_{PNE}$, we may derive a PQE set $\mathcal{A}_\epsilon (\mathbf{a}_{PNE}) \subseteq \mathcal{A}_\epsilon$ with better performance by "equal response" move. The equal response move refers to that, one user changes its connected network with the QoE level maintained, resulting in a PQE where at least one user's QoE is improved and other users' QoE remain unchanged. If $\mathcal{A}_\epsilon (\mathbf{a}_{PNE}) \neq \emptyset$, then $\forall \mathbf{a} \in \mathcal{A}_\epsilon (\mathbf{a}_{PNE})$, $\mathbf{a} \triangleright \triangleright \mathbf{a}_{PNE}$.

One MQE, we can obtain the following result.

Proposition 7.2 *Every finite QoE game* G_{QoE} *possesses MQE.*

Proof We define a new game $G' = \left(\mathcal{M}, \{\mathcal{A}_m\}_{m \in \mathcal{M}}, \{u_m (\mathbf{a})\}_{m \in \mathcal{M}} \right)$, where the user utility function is defined based on QoE weight $u_m (\boldsymbol{a}) = x_{m,k}$, $\boldsymbol{a} \in Q_{m,k}$, $\forall \boldsymbol{a}$, $\forall m$.

Since G' is a finite game, it possesses mixed strategy Nash equilibrium [12]. Denote the mixed Nash equilibrium of G' by π'. According to the definition of mixed strategy Nash equilibrium, for $\forall m$ and given π_{-m}, $\forall \pi_m \in \Pi_m$,

$$\sum_{\mathbf{a} \in \mathcal{A}_m} u_m (\mathbf{a}) P \left\{ \mathbf{a} \,\middle|\, \pi' \right\} \geq \sum_{\mathbf{a} \in \mathcal{A}_m} u_m (\mathbf{a}) P \left\{ \mathbf{a} \,\middle|\, \left(\pi_m, \pi'_{-m} \right) \right\} \tag{7.8}$$

where $P \{\mathbf{a} \,|\, \pi\}$ denotes the probability of achieving pure strategy $\mathbf{a}$ given π. According to the definition of $u_m (\mathbf{a})$,

$$\sum_{\mathbf{a} \in \mathcal{A}_m} u_m (\mathbf{a}) P (\mathbf{a} \,|\, \pi) = \sum_{\mathbf{a} \in Q_{m,k}} u_m (\mathbf{a}) P (\mathbf{a} \,|\, \pi) = \sum_k x_{m,k} p_{m,k} (\pi)$$

Thus, (7.8) is equivalent to

$$\sum_k x_{m,k} p_{m,k} \left(\pi' \right) \geq \sum_k x_{m,k} p_{m,k} \left(\pi_m, \pi'_{-m} \right)$$

which is in line with the definition of MQE. Hence, π' is also a MQE of the QoE game. $\qquad\qquad\square$

7.4.3 User Demand Diversity Gain

Fortunately, we can prove that there exists PQE for the considered network selection game. The first theorem is suitable for a special weight setting.

Theorem 7.2 *The considered QoE game has at least one PQE, if $\forall n \in N$, $\forall m, m' \in M$, $w_{m,n} = w_{m',n}$.*

Proof When $\forall n \in N$, $\forall m, m' \in M$, $w_{m,n} = w_{m',n}$, the QoS game $\left(M, \{\mathcal{A}_m\}_{m \in M}, \{\theta_m\}_{m \in M}\right)$ is reduced to a weighted congestion game in [13], which has been proved to have a pure strategy Nash equilibrium. According to Proposition 2, the corresponding QoE game also has a PQE. $\qquad\square$

The following theorem reveals that PQE exists when each user's data rates to different networks are comparable.

Theorem 7.3 *The QoE game G_{QoE} has at least one PQE if there exists some states (specified in the following proof) satisfying $\forall n$, $W_n > \max\limits_{\phi \in \Phi_n} \dfrac{-\log \phi}{1-\phi}$, where*

$$\Phi_n = \left\{ \phi \mid R_{m,j'} = \phi R_{m,j}, \forall m \in M_n, j, j' \in \mathcal{A}_m \; j \neq j' \right\}.$$

Proof We define other two games. The first game is $G_1 = \left(M, \{\mathcal{A}_m\}_{m \in M}, \{r_m\}_{m \in M}\right)$, where each user behaves to minimize the cost $r_m = \dfrac{e^{W_n}}{w_m R_{m,n}}$. This game belongs to weighted congestion games with player-specific utility, which has no general result on the existence of equilibriums. The second game is $G_2 = \left(M, \{\mathcal{A}_m\}_{m \in M}, \{\zeta_m\}_{m \in M}\right)$, where each user tries to minimize the cost $\zeta_m = \dfrac{W_n}{w_m R_{m,n}}$. Denote the PNE sets of G_1 and G_2 as $\mathcal{A}^1_{PNE}$ and $\mathcal{A}^2_{PNE}$, respectively. The "some states" in Theorem 7.3 refer to a subset of PNE in G_1, which is denoted as $\tilde{\mathcal{A}}^1_{PNE} \subseteq \mathcal{A}^1_{PNE}$. The proof relies on the relationship $\tilde{\mathcal{A}}^1_{PNE} \subseteq \mathcal{A}^2_{PNE} = \mathcal{A}_{PNE} \subseteq \mathcal{A}_{PQE}$. We prove Theorem 7.3 by demonstrating $\mathcal{A}^1_{PNE} \neq \emptyset$ and validating the above relationship. In the following, Lemma 7.1 shows that $\mathcal{A}^1_{PNE} \neq \emptyset$. Lemma 7.2 validates that there exists a subset of PNE in G_1 belonging to $\mathcal{A}^2_{PNE}$, that is, $\tilde{\mathcal{A}}^1_{PNE} \subseteq \mathcal{A}^2_{PNE}$, if the condition in Theorem 7.3 is met for $\forall \mathbf{a} \in \tilde{\mathcal{A}}^1_{PNE}$.

Lemma 7.1 *Game G_1 possesses at least one PNE.* $\qquad\square$

Proof Since $\log r_m = W_n - \log R_{m,n} - \log w_m$ and w_m is fixed, minimizing the original r_m is equivalent to minimizing a redefined $r_m = W_n - \log R_{m,n}$. We define a function $\Upsilon : \mathbf{a} \to \mathfrak{R}$:

$$\Upsilon(\mathbf{a}) = \sum_{m \in M} w_m \left(-2 \log R_{m,a_m} + W_{a_m} + w_m \right)$$

For any user $l \in M$, suppose that it improves r_l by unilaterally deviating its strategy from s to t. Denote the sets of users associated with network s and t before user l deviating its strategy as M_s and M_t, the corresponding user weights in the networks are $W_s = \sum\limits_{m \in \mathcal{M}_s} w_m$ and $W_t = \sum\limits_{m \in \mathcal{M}_t} w_m$, respectively. The change in the cost r_l is derived as

$$
\begin{aligned}
r_l\left(t, \mathbf{a}_{-l}\right) &- r_l\left(s, \mathbf{a}_{-l}\right) \\
&= W_t + w_l - \log R_{l,t} - W_s + \log R_{l,s} \\
&< 0.
\end{aligned}
\tag{7.9}
$$

At the same time, the changes in Υ are shown in (7.10).

$$
\begin{aligned}
\Upsilon\left(t, \mathbf{a}_{-l}\right) &- \Upsilon\left(s, \mathbf{a}_{-l}\right) \\
&= \sum_{m \in \mathcal{M}_t} w_m \left(-2 \log R_{m,t} + (W_t + w_l) + w_m\right) \\
&+ \sum_{m \in \mathcal{M}_s, m \neq l} w_m \left(-2 \log R_{m,s} + (W_s - w_l) + w_m\right) \\
&- \sum_{m \in \mathcal{M}_t} w_m \left(-2 \log R_{m,t} + W_t + w_m\right) - \sum_{m \in \mathcal{M}_s, m \neq l} w_m \left(-2 \log R_{m,s} + W_s + w_m\right) \\
&+ w_l \left(-2 \log R_{l,t} + (W_t + w_l) + w_l\right) - w_l \left(-2 \log R_{l,s} + W_s + w_l\right).
\end{aligned}
\tag{7.10}
$$

Note that the first four terms represent the changes introduced by users associated with network s and t except for user l, the remaining terms are the changes introduced by user l. In the first four terms, only the total user weight of each network is changed, thus they can be reduced to $\sum\limits_{m \in \mathcal{M}_t} w_m w_l - \sum\limits_{m \in \mathcal{M}_s, m \neq l} w_m w_l$. We derive the resulting change in Υ as

$$
\begin{aligned}
\Upsilon\left(t, \mathbf{a}_{-l}\right) &- \Upsilon\left(s, \mathbf{a}_{-l}\right) \\
&= w_l \left(-2 \log R_{l,t} + W_t + 2w_l\right) - w_l \left(-2 \log R_{l,s} + W_s + w_l\right) \\
&+ w_l \sum_{m \in \mathcal{M}_t} w_m - w_l \sum_{m \in \mathcal{M}_s, m \neq l} w_m \\
&= w_l \left(-2 \log R_{l,t} + W_t + 2w_l\right) - w_l \left(-2 \log R_{l,s} + W_s + w_l\right) \\
&+ w_l W_t - w_l \left(W_s - w_l\right) \\
&= 2w_l \left(-\log R_{l,t} + \log R_{l,s} + W_t - W_s + w_l\right) \\
&= 2w_l \left[r_l\left(t, \mathbf{a}_{-l}\right) - r_l\left(s, \mathbf{a}_{-l}\right)\right] \\
&< 0.
\end{aligned}
\tag{7.11}
$$

The results in (7.9) and (7.11) imply that the changes in Υ and r_l possess the same sign. Therefore, G_1 is an ordinary potential game, where $\Upsilon(\mathbf{a})$ is the potential function. Since G_1 is finite, it possesses at least one PNE [12]. This game is a special case of congestion games in [14].   □

Lemma 7.2 *For games G_1 and G_2, denote a PNE subset of G_1 as $\tilde{\mathcal{A}}_{PNE}^1 \subseteq \mathcal{A}_{PNE}^1$, if the following condition holds $\forall \mathbf{a} \in \tilde{\mathcal{A}}_{PNE}^1$, it satisfies*

$$W_n > \max_{\phi \in \Phi_n} \frac{-\log \phi}{1 - \phi}, \forall n \in \mathcal{N}$$

where $\Phi_n = \left\{ \phi \,\middle|\, R_{m,j'} = \phi R_{m,j}, \forall m \in \mathcal{M}_n, j, j' \in \mathcal{A}_m \ j \neq j' \right\}$, *we can get* $\tilde{\mathcal{A}}_{PNE}^1 \subseteq \mathcal{A}_{PNE}^2$. $\qquad\square$

Proof Denote a PNE of G_1 as $\mathbf{a}^*$, according to the definition of PNE, we know that for any user $m \in \mathcal{M}$, given the other users' strategy fixed as $\mathbf{a}_{-m}^*$, if it unilaterally deviates the strategy from $a_m = j$ to $a_m = j'$, we have

$$\frac{e^{W_{j'} + w_m}}{w_m R_{m,j'}} \geq \frac{e^{W_j}}{w_m R_{m,j}},$$

where W_j and $W_{j'}$ are the total weights of networks j and j' before m deviating its strategy, respectively. Thus,

$$\log R_{m,j} + W_{j'} + w_m \geq \log R_{m,j'} + W_j,$$

And then

$$W_{j'} + w_m \geq \log R_{m,j'} - \log R_{m,j} + W_j.$$

Define $R_{m,j'} = \phi R_{m,j}, j, j' \in \mathcal{A}_m$, we study the property of strategy $\mathbf{a}^*$ in game G_2. Given other users' strategy fixed as $\mathbf{a}_{-m}^*$, the reward change of user m by unilaterally deviating its strategy from $a_m = j$ to $a_m = j'$ is

$$
\begin{aligned}
\frac{W_{j'} + w_m}{w_m R_{m,j'}} - \frac{W_j}{w_m R_{m,j}} &\geq \frac{\log R_{m,j'} - \log R_{m,j} + W_j}{w_m R_{m,j'}} - \frac{W_j}{w_m R_{m,j}} \\
&= \frac{\log \phi + W_j}{w_m \phi R_{m,j}} - \frac{W_j}{w_m R_{m,j}} \\
&= \frac{\log \phi + W_j - \phi W_j}{w_m \phi R_{m,j}}
\end{aligned}
\tag{7.12}
$$

It is easy to find that if $W_j > \frac{-\log \phi}{1 - \phi}$, that is, $\log \phi + W_j - \phi W_j > 0$, then $\frac{W_{j'} + w_m}{w_m R_{m,j'}} \geq \frac{W_j}{w_m R_{m,j}}$, which indicates that $\mathbf{a}^*$ is also a PNE of G_2. $\qquad\square$

Moreover, maximizing $\frac{w_m R_{m,n}}{W_n}$ in G_θ is equivalent to minimizing $\frac{W_n}{w_m R_{m,n}}$ in G_2, hence $\mathcal{A}_{PNE}^2 = \mathcal{A}_{PNE}$. In addition, Theorem 1 ensures $\mathcal{A}_{PNE} \subseteq \mathcal{A}_{PQE}$. Combining the above results, we find that $\mathbf{a}^*$ is a PQE of G_{QoE}, thus $\tilde{\mathcal{A}}_{PNE}^1 \subseteq \mathcal{A}_{PQE}$. $\qquad\square$

It is worth noting that the condition in Theorem 2 is considerably mild, since it essentially requires that the data rate gaps ϕ among different wireless networks are finite. Actually, the date rates $R_{m,n}$ to different wireless networks n are constrained. For example, date rates in 4G (100 Mbps in LTE and WiMax) or foreseeable 5G cellular networks are comparable to new emerging WLAN standards such as IEEE 802.11 n/ac/ad (200 to 1G Mbps). We assume a maximal 20-fold gap between any two networks, i.e., $0.05 \leq \phi \leq 20$, $\frac{-\log \phi}{1-\phi} < 3.15$. If we ensure that each user weight is larger than 1, it is natural to get $W_n > 3.15$ considering that each network at least has several users connected. Therefore, the PQE exists in most cases.

Having ensured the existence of PQE, we can formally define user demand diversity gain based on Theorem 7.1 and Corollary 7.1.

Definition 7.7　Given the distributed RRM problem above, the user demand diversity gain η is the system-level QoE gap between the best PQE $\mathbf{a}^*_{PQE}$ and the best PNE $\mathbf{a}^*_{PNE}$

$$\eta = \mathbf{D}\left(\mathbf{a}^*_{PQE}\right) - \mathbf{D}\left(\mathbf{a}^*_{PNE}.\right) \tag{7.13}$$

As can be seen, the user demand diversity gain is defined based on equilibriums in the QoE game and the throughput game. In this way, it captures the distributed decision-making property and highlights the effect of user demand diversity. Clearly, if $\mathbf{a}^*_{PQE} \rhd \mathbf{a}^*_{PNE}$, there is at least one positive gain at some level. If $\mathbf{a}^*_{PQE} \rhd \rhd \mathbf{a}^*_{PNE}$, there is no negative gain and at least one positive gain at all levels.

7.5　MARL Algorithms for QoE Game

In this section, two game-theoretic MARL algorithms will be designed. The first one is a stochastic learning based algorithm that is able to converge to PQEs of the QoE game. The second one could successfully exploit the user demand diversity gain by achieving the best PQEs.

7.5.1　Stochastic Learning Automata Based QoE Equilibrium Learning Algorithm

We propose a QoE equilibrium learning algorithm as shown in Algorithm 1. The idea of this algorithm is transforming the QoE equilibrium learning to Nash equilibrium learning by normalized reward. In the algorithm, $0 < b < 1$ is a parameter, the normalization function $g(\cdot)$ satisfies

- $0 \leq g(q_m(\mathbf{a})) \leq 1$ for $\forall \mathbf{a} \in \mathcal{A}, \forall m \in \mathcal{M}$.
- $\forall j$, if $\mathbf{a}, \mathbf{a}' \in Q_{m,j}$, then $g(q_m(\mathbf{a})) = g(q_m(\mathbf{a}'))$.
- $\mathbf{a} \in Q_{m,j}$ and $\mathbf{a}' \in Q_{m,k}$, for $\forall j, k$, if $j \leq k$, then $g(q_m(\mathbf{a})) \geq g(q_m(\mathbf{a}'))$.

The proposed algorithm is a modified stochastic learning automata (SLA) in [15]. The users only have to know the reward in each iteration and the normalization function for the algorithm running. The algorithm could be run in an online manner, that is, the user keeps a duration of communication after each decision, the slot is relatively large (several seconds), accordingly. Such online manner could effectively relieve the frequent handoffs and delay.

We present the convergence result in Theorem 7.4 without proof, since it follows the similar way of the potential game problem in [16]. It is worth noting that even the condition in the proposition is not met, the proposed algorithm can still converge to PQE in most cases according to our simulation results.

Theorem 7.4 *When $R_{m,n} = R_{m',n}$, $w_{m,n} = w_{m',n}$, $\forall m, m' \in \mathcal{M}$, $n \in \mathcal{N}$, the QoE equilibrium learning algorithm converges to a PQE.*

Algorithm 10 SLA-based PQE Learning Algorithm

1: **Initiate:** $\forall n \in \mathcal{N}$, $\forall m \in \mathcal{M}$, $\pi_{m,n}(0) = \frac{1}{|\mathcal{A}_m|}$.
2: **loop**
3:     Each user m selects the network according to the following rules:
4:     **if** $\mathbf{a}(t-1) \in Q_{m,1}$ **then**
5:         $a_m(t) = a_m(t-1)$
6:     **else**
7:         $a_m(t)$ is selected according to the strategy distribution probability $\pi_m(t)$.
8:     **end if**
9:     Each user gets a normalized reward $\hat{q}_m = g(q_m(\mathbf{a}(t)))$ and updates π_m according to the following rule
10:     **if** $n = a_m(t)$ **then**
11:         $\pi_{m,n}(t+1) = \pi_{m,n}(t) + b\hat{q}_m(t)\left(1 - \pi_{m,n}(t)\right)$
12:     **else**
13:         $\pi_{m,n}(t+1) = \pi_{m,n}(t) - b\hat{q}_m(t)\,\pi_{m,n}(t)$.
14:     **end if**
15: **end loop**

7.5.2 Trial and Error Based QoE Equilibrium Refinement Algorithm

Note that SLA-based PQE learning algorithm only converges to PQE, but not the best PQE, which indicates that the achieved performance gain over PNE is limited. In order to fully exploit the user demand diversity, it is necessary to enforce the system to efficient PQEs.

To get around the above challenge, our idea is reassigning reward to QoE levels and transforming the PQE learning to equivalent PNE learning problem. To this end, we define a reassignment function $g(\cdot)$ to map QoE levels to quantifiable reward $u_m = g(q_m(\mathbf{a}))$. The reassignment function must satisfy the following Condition 1.

Condition 1:

- $g(q_m(\mathbf{a})) \geq 0, \forall \mathbf{a} \in \mathcal{A}$;
- $g(q_m(\mathbf{a})) = g(q_m(\mathbf{a}')) = c_{\kappa_m(\mathbf{a})}, \forall \mathbf{a}, \mathbf{a}' \in \mathcal{A}$ with $\kappa_m(\mathbf{a}) = \kappa_m(\mathbf{a}')$, where c_j, $1 \leq j \leq 5$ is a specific constant for QoE level j;
- $c_j < c_{j+1}, \forall 1 \leq j < 4$.

The first sub-condition restricts the reward to be nonnegative. The second subcondition requires that strategies with the same QoE level result in the same reward. The third sub-condition ensures that a better QoE level results in a larger reward.

Based on the reassigned reward, we derive the QoE equilibrium learning algorithm by resorting to the trail and error (TE) learning algorithm in [17]. This learning algorithm possesses several desired advantages. First, it is fully distributed without relying on information exchanges among users. Second, it is possible to get an efficient PNE in terms of the social welfare. Third, it has no restriction on users' utility form and only relies on individual's reward feedback. In the trail and error learning, a user m's state is represented by a triplet $(mood_m, \bar{a}_m, \bar{u}_m)$, where $mood_m$ is user's mood which indicates user's expectation state and determines user's strategy search rule, $\bar{a}_m$ and $\bar{u}_m$ are its benchmark strategy and benchmark reward, respectively. Users' mood can be in four states, "Content" (C), "Hopeful" (C+), "Watchful" (C-), and "Discontent" (D). The state transition diagram of the algorithm for each user is shown in Fig. 7.2.

We now detail the updating strategy and the state transition condition among states. Denote the selected strategy and realized reward at current time instant by a_m and u_m. The algorithm operates in an iterative manner, where strategy selection is determined by the triplet state $(mood_m, \bar{a}_m, \bar{u}_m)$ as follows.

- *In the Content state* $(C, \bar{a}_m, \bar{u}_m)$: The user experiments or does not experiment with probability ε and $1 - \varepsilon$. When an experiment happens, the strategy a_m is selected according to a uniform random distribution from the set $\mathcal{A}_m \backslash \bar{a}_m$ and the state updating rule is: (1) If $u_m > \bar{u}_m$, with probability $\varepsilon^{G(u_m - \bar{u}_m)}$, it changes to the new benchmark strategy and reward, i.e., $(C, \bar{a}_m, \bar{u}_m) \rightarrow (C, a_m, u_m)$ (transition "a"); with probability $1 - \varepsilon^{G(u_m - \bar{u}_m)}$, the state remains unchanged (transition "a"). We will discuss the function $G(\cdot)$ later. (2) If $u_m \leq \bar{u}_m$, the state remains unchanged (transition "a"). On the other hand, when the user does not experiment, the benchmark strategy $a_m = \bar{a}_m$ is selected and the updating rule is: (1) If $u_m > \bar{u}_m$, only the mood changes to $C+$ (transition "b"), i.e., $(C, \bar{a}_m, \bar{u}_m) \rightarrow (C+, \bar{a}_m, \bar{u}_m)$. (2) If $u_m < \bar{u}_m$, only the mood changes to $C-$ (transition "c"), i.e., $(C, \bar{a}_m, \bar{u}_m) \rightarrow (C-, \bar{a}_m, \bar{u}_m)$. (3) Otherwise, the state remains unchanged (transition "a").
- *In the Discontent state* $(D, \bar{a}_m, \bar{u}_m)$: The strategy a_m is selected randomly according to the uniform distribution from the strategy set $\mathcal{A}_m$. The state transits to (C, a_m, u_m) (transition "i") with probability $\varepsilon^{F(u_m)}$ and remains unchanged (transition "j") with probability $1 - \varepsilon^{F(u_m)}$. We will also discuss $F(\cdot)$ later.
- *In the Hopeful state* $(C+, \bar{a}_m, \bar{u}_m)$: The benchmark strategy $a_m = \bar{a}_m$ is selected. (1) If $u_m > \bar{u}_m$, only the mood and benchmark reward change (transition "d"), i.e., $(C+, \bar{a}_m, \bar{u}_m) \rightarrow (C, \bar{a}_m, u_m)$. (2) Similarly, if $u_m < \bar{u}_m$, only the mood and

Fig. 7.2 State evolution of trail and error algorithm

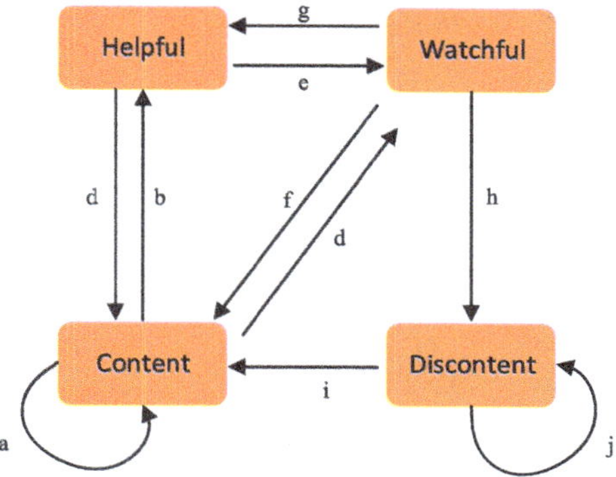

benchmark reward change (transition "e"), i.e., $(C+, \bar{a}_m, \bar{u}_m) \to (C-, \bar{a}_m, u_m)$. (3) If $u_m = \bar{u}_m$, only the mood changes and the state transits to $(C, \bar{a}_m, u_m)$ (transition "d").

- *In the Watchful state* $(C-, \bar{a}_m, \bar{u}_m)$: The benchmark strategy $a_m = \bar{a}_m$ is selected and only the mood changes as: (1) If $u_m > \bar{u}_m$, the state transits to $(C+, \bar{a}_m, \bar{u}_m)$ (transition "g"). (2) If $u_m < \bar{u}_m$, the state transits to $(D, \bar{a}_m, \bar{u}_m)$ (transition "h"). (3) Otherwise, the state transits to $(C, \bar{a}_m, \bar{u}_m)$ (transition "f").

In the algorithm, $G\,(\cdot)$ and $F\,(\cdot)$ control some state transition probabilities. Following a general setting in [17], they are the following forms:

$$F\,(u_m) = \alpha_1 u_m + \beta_1, \alpha_1 < 0$$
$$G\,(u_m - \bar{u}_m) = \alpha_2\,(u_m - \bar{u}_m) + \beta_2, \alpha_2 < 0$$

The parameters α and β are set to bound the two probabilities

$$0 < F\,(u_m) < \frac{1}{2M} \tag{7.14}$$

$$0 < G\,(u_m - \bar{u}_m) < \frac{1}{2} \tag{7.15}$$

Although the original trail and error algorithm can converge to an efficient PNE of a game with the maximal social welfare, it is generally not the efficient PQE. The reason is that the strategy $\mathbf{a}^* = \arg\max\limits_{\mathbf{a} \in \mathcal{A}} \sum\limits_{m \in M} u_m\,(\mathbf{a})$ is not necessarily the same as the efficient PQE in (7.6). However, we found that, by imposing some restrictions on the reassignment function, the efficient PQE is attainable.

Theorem 7.5 *The QoE equilibrium learning algorithm converges to the efficient PQE of G_{QoE} if the reassignment function satisfies Condition 1 and the following Condition 2.*

Condition 2: *For $i = 3, 4, 5$,*

$$\frac{c_i - c_{i-1}}{c_{i-1} - c_{i-2}} > M - 1.$$

Proof After the reward reassignment, we can get a new game

$$G_3 = \left(\mathcal{M}, \{\mathcal{A}_m\}_{m \in \mathcal{M}}, \{u_m\}_{m \in \mathcal{M}} \right)$$

and denote its PNE set as $\mathcal{A}^3_{PNE}$. We notice that the preference order of G_{QoE} is the same as that of G_3 provided that Condition 1 is satisfied, thus the equilibrium sets are the same, i.e., $\mathcal{A}^3_{PNE} = \mathcal{A}_{PQE}$.

The second step is to demonstrate that Condition 2 guarantees that the efficient PQE in G_{QoE} corresponds with the best PNE (maximizing the sum reward of users) in G_3. Denote the set of efficient PQE as

$$\mathcal{A}^*_{PQE} = \left\{ \mathbf{a}^* \rhd \mathbf{a} \,\middle|\, \mathbf{a}^* \rhd \mathbf{a} \text{ or } \mathbf{a}^* \doteq \mathbf{a}, \forall \mathbf{a} \in \mathcal{A}_{PQE} \right\}$$

Given the efficient PQE $\mathbf{a}^* \in \mathcal{A}^*_{PQE}$ and any $\mathbf{a}' \in \mathcal{A}_{PQE} \backslash \mathcal{A}^*_{PQE}$, the gap of the sum reward between these two strategy profiles is

$$\sum_{m \in \mathcal{M}} u_m \left(\mathbf{a}^* \right) - \sum_{m \in \mathcal{M}} u_m \left(\mathbf{a}' \right) \tag{7.16}$$

$$= \sum_{i=1}^{4} c_i \left\{ \left[d_i \left(\mathbf{a}^* \right) - d_{i+1} \left(\mathbf{a}^* \right) \right] - \left[d_i \left(\mathbf{a}' \right) - d_{i+1} \left(\mathbf{a}' \right) \right] \right\}$$

$$+ c_5 \left[d_5 \left(\mathbf{a}^* \right) - d_5 \left(\mathbf{a}' \right) \right]$$

$$= \sum_{i=2}^{5} \left[c_i - c_{i-1} \right] \left[d_i \left(\mathbf{a}^* \right) - d_i \left(\mathbf{a}' \right) \right]$$

According to the definition of the efficient strategy profile in (7.6), there exists $1 < j \le 5$ satisfying $d_i \left(\mathbf{a}^* \right) = d_i \left(\mathbf{a}' \right)$ for $j < i \le 5$ and $d_j \left(\mathbf{a}^* \right) \ge d_j \left(\mathbf{a}' \right) + 1$. Moreover,

$$\sum_{i=2}^{j-1} \left[d_i \left(\mathbf{a}^* \right) - d_i \left(\mathbf{a}' \right) \right] \ge 1 - M,$$

thus

$$\sum_{i=2}^{5} \left[c_i - c_{i-1} \right] \left[d_i \left(\mathbf{a}^* \right) - d_i \left(\mathbf{a}' \right) \right] \ge c_j - c_{j-1} + (1 - M) \max_{i=2,\dots,j-1} \left(c_i - c_{i-1} \right)$$

$$\tag{7.17}$$

Therefore, if $\forall i = 3, 4, 5$,

$$\frac{c_i - c_{i-1}}{c_{i-1} - c_{i-2}} > M - 1$$

holds, we get

$$\sum_{m \in \mathcal{M}} u_m \left(\mathbf{a}^* \right) - \sum_{m \in \mathcal{M}} u_m \left(\mathbf{a}' \right) > 0.$$

That is, the efficient PQE in G_{QoE} is also the best PNE in G_3.

At the third step, we know that the best PNEs in G_3 are the stochastically stable states of the trail and error algorithm [17]. Combining above three steps completes the proof. $\square$

Note that the Condition 2 can be regarded as a design rule from the system-level perspective, that is, the system designer would like to impose this condition on each user's learning algorithm to enforce the system to arrive at desired states. Different from the convergence concept in many existing learning algorithms, the convergence here means that given any small α, there exists a ϵ_α such that if $0 < \varepsilon < \varepsilon_\alpha$, the best PQEs arise with at least $(1 - \alpha)$ fraction of time.

A common limitation of most existing distributed learning algorithms (e.g., [18]), including the proposed QoE equilibrium learning algorithm, is the low convergence speed, which may incur considerable cost. In the existing distributed learning framework, each user generally works in a three-step cycle, "Learning-Decision-Action". That is, each user updates some variables and processes following some predefined learning rules in "Learning" stage. Based on the updated results, the user further makes a new decision at the "Decision" stage according to the input variables. Finally, the decision result is executed in the "Action" stage, followed by a new cycle. Such type of learning algorithms can be found in [18]. The QoE equilibrium learning algorithm falls into this type. As mentioned above, the overloaded network handoff cost in this framework severely hinders the QoE equilibrium learning algorithm's implementation.

Considering the emerging convergence of cloud computing and wireless communications, we propose a cloud-assisted learning framework to accommodate the PQE learning algorithm with reduced cost and faster convergence speed. We summarize the framework in Fig. 7.3. Compared with the existing framework, users do not need to execute the network handoff in our framework. Instead, they only report the decision result to the cloud. A load aggregator is responsible for gathering all users' decisions and makes corresponding virtual load information distribution process. Different from the centralized scheme in [19], the load aggregator cloud does *not* make the RRM decision. The proposed framework can be applied to [3, 7, 18]. Some mobility-related functions can also be delegated to the cloud.

Depending on the trust level, we can imagine two modes for the algorithm in the framework, full trust mode and partial trust mode. In full trust mode, users are willing to report all necessary information to the cloud including the date rate set $R_m = \left\{ R_{m,n} \,|\, n \in \mathcal{A}_m \right\}$ and the demand information Q_m. On behalf of the users, the load aggregator cloud gathers all users' information and runs the QoE learning algorithm locally. On the other hand, in the partial trust mode, users may not be willing to report R_m and Q_m due to security and privacy considerations. In this case, the users and the load aggregator cloud work in an interactive manner. The procedures of the full trust mode and the partial trust mode are shown in the tables. Note that the convergence property of the QoE learning algorithm still holds in this framework. In the original algorithm and similar algorithms in [3, 18], one iteration may involve a network handoff process as well as a transmission duration to explore the network load information. While even in the partial trust mode of the cloud-assisted learning, one iteration only corresponds to one message exchange in a much smaller time scale. The algorithm can run on the background, with limited influence on normal communications. Compared with the traditional implementation, the convergence time and cost in the cloud-assisted learning framework is significantly reduced owing to its refraining from handoff.

Algorithm 11 Full trust mode for TE-based PQE refinement algorithm

---

1: **User Side**:
2:    Access one of the networks and register in the load aggregator cloud. Report R_m and Q_m to the cloud.
3:    Wait for the cloud to return suggested network access decision a_m.
4: **On the Load Aggregator Cloud**:
5:    Receive users' R_m and Q_m, maintain accumulative decision distribution $U_m = (num_1, ..., num_{|\mathcal{A}_m|})$ for each user m, where num_n, $1 \le n \le |\mathcal{A}_m|$ denotes the number of times user m selecting network n. Run the QoE learning algorithm locally and update U_m, until some stop rule is met.
6:    Return suggestions on network selection to users.

Algorithm 12 Partial trust mode for TE-based PQE refinement algorithm

---

1: **User Side**:
2:    Access one of the networks and register in the load aggregator cloud.
3: **loop**
4:    Make the network selection decision $a_m(t)$ according to the TE- based learning algorithm.
5:    Report $a_m(t)$ to the cloud.
6:    Wait for the cloud to return the virtual load $W_{a_m(t)}(t)$. Calculate the quantifiable reward $u_m(t)$, update the learning algorithm.
7:    Execute a final network selection until some stop rule is met.
8: **end loop**
9: **On the Cloud**:
10: **loop**
11:    Gather users' decisions and calculate each networks' virtual load $W_n(t)$.
12:    Return each user m the corresponding network virtual load $W_{a_m(t)}(t)$.
13:    Stop until some stop rule is met.
14: **end loop**

Fig. 7.3 On cloud learning framework

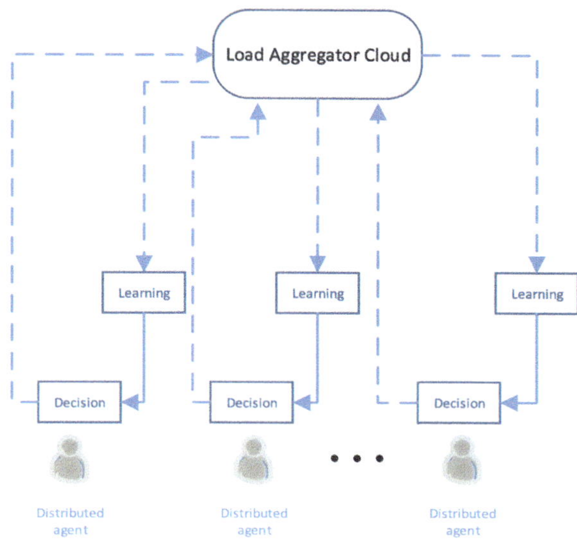

7.6 Simulation Results

In this section, some simulation results are presented.

7.6.1 Simulation Setting

In our simulation, a heterogeneous wireless network deployment scenario in LTE-A is considered. One picocell locates at the partially overlapping area of the two macrocells and the other picocell locates at the coverage area of one macrocell. The simulation parameters of the networks are set according to [20]. Specifically, the path loss from a macrocell and a picocell to the user are $L = 128.1 + 37.6\log_{10}R$ and $L = 140.7 + 36.7\log_{10}R$, where R is the distance between the user and the base station in km. The macrocell radius and the picocell radius are 500 and 100 m. The standard deviation of lognormal shadowing is 10 dB. The base station transmitting powers of the macrocell and the picocell are 46 dBm in 20 MHz carrier and 30 dBm in 10 MHz carrier, respectively. We distinguish the background load and the dynamic load. The background load, denoted by M', refers to the total number of users who have only one network available. The dynamic load refers to the number of users who locate at overlapping areas of networks and thus can select their access networks freely, which is exactly the user number M in the system model. Denote the user weights in (7.1) for group video calling user, HD video calling user, and general video calling user as w_1, w_2 and w_3, $\mathbf{W} = (w_1, w_2, w_3)$, respectively. We further consider two cases of user weight: equal weights $\mathbf{W} = (1, 1, 1)$ and unequal weights $\mathbf{W} = (4, 3, 1)$.

We take three types of video calling users in Skype as examples of user demand model. The first type is the group video calling user with the required minimal throughput 512 kbps and the recommended throughput 2 Mbps. The second type is the high definition (HD) video calling user with the required minimal throughput 1.2 Mbps and the recommended throughput 1.5 Mbps. The third type is the general video calling user with the required minimal throughput 128 kbps and the recommended throughput 500 kbps. The user-type classification and throughput requirements follow the recommendations from Skype [21]. We let the output MOS at the minimal throughput and the recommended throughput to be 2 and 4, respectively. We consider that the minimal throughput only supports the basic user demand (level "Poor"), while the recommended throughput is supposed to offer sufficiently good user experience (level "Good"). Based on these two MOS levels, we can obtain the explicit QoE function of each user type. Users are randomly distributed in the networks. Each user falls into one of the three types with equal probabilities.

7.6.2 Results

In this section, the existence of user demand diversity gain is verified. Then, the proposed two MARL algorithms will be simulated and validated.

7.6.2.1 User Demand Diversity Gain

Since the actual user demand distributions can be different, we simulate the user demand diversity gains under different user-type distributions. We fix $M' = 60$, $M = 20$ and generate different user locations and user-type distributions. Four cases of user-type distributions are considered. In the first case, all 20 users are the HD video calling user. In the second case, 10 users are the group video calling user and 10 are the HD video calling user. In the third case, all users are the group video calling user. In the fourth case, the numbers of general video calling users, the HD video calling users, and the group video calling users are 7, 7, and 6, respectively. We randomly generate 50 different user location distributions to see PNE and PQE numbers. It is found that the number of PNE is around 10, while the number of PQE is much larger, several thousands. We then compare the average user demand diversity gains in Figs. 7.4 and 7.5. It is exciting to observe that there are user demand diversity gain at all cases. Especially, the gain at the "Excellent" level in the third case is the largest of all, i.e., larger than 2. Even when all users have the same demand type, there still exist performance gains. These results imply that the user demand gain is impacted by the user-type distribution and not limited to the cases with diverse user demands.

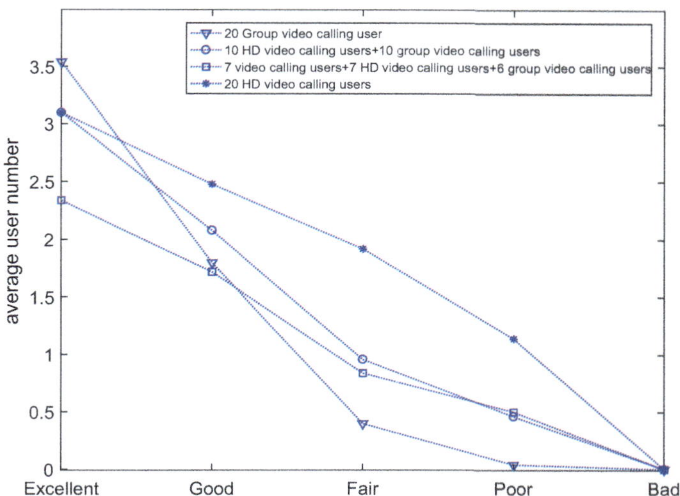

Fig. 7.4 User demand diversity gain with equal user weights

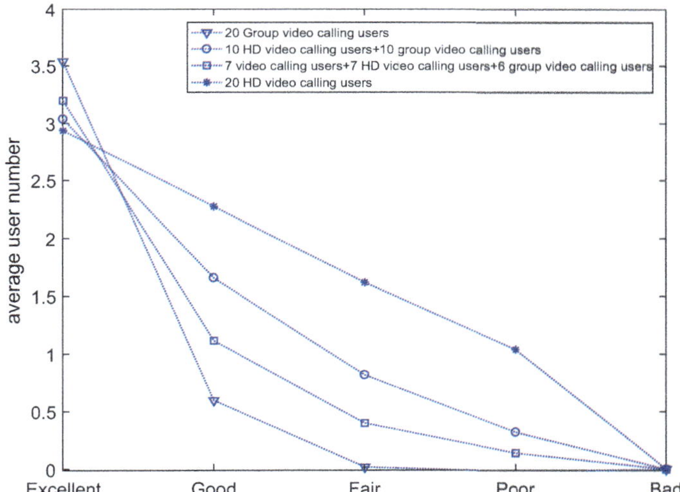

Fig. 7.5 User demand diversity gain with unequal user weights

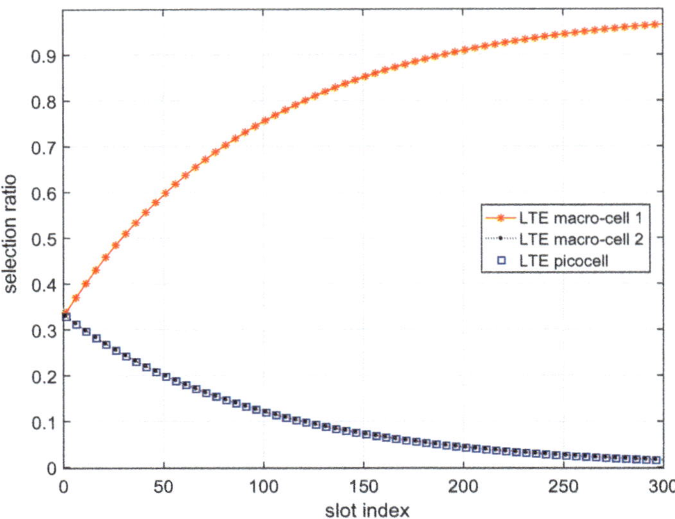

Fig. 7.6 The accumulative network selection probability distribution in a sample run of the proposed algorithm

7.6.2.2 SLA-Based PQE Learning Algorithm

Consider a scenario with 15 users in overlapping areas of the networks, where the number of general video calling users, the HD video calling users, and the group video calling users are 6, 5, and 4, respectively. To validate the proposed SLA-based PQE learning algorithm, we show the network selection probability distribution sample of one user in the overlapping area of the two LTE macrocells and the picocell in Fig. 7.6. It is seen that the algorithm gradually converges to sticking to one of the LTE macrocell, which means that the algorithm is able to converge to pure strategies.

We also apply the algorithm in the above scenario with different iteration numbers and learning parameter b. As can be seen in Fig. 7.7, the ratio of converging to a PQE grows with the increase of b and the iteration number. We further compare the proposed algorithm with another two equilibrium learning algorithms. One is the throughput maximization algorithm in [3], where at each iteration, users change their access network to get a larger throughput. The other one is the satisfaction equilibrium learning algorithm in [8]. The system-level QoE at the 1000 iteration averaged by 500 runs are shown in Fig. 7.8. For QoE levels better than "Poor", the proposed algorithm always has a larger number of users. We can conclude that the proposed algorithm outperforms the other two algorithms in improving users' QoE.

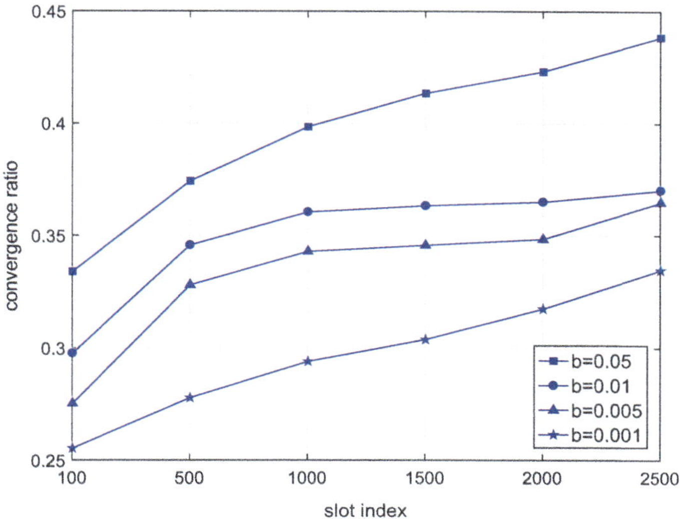

Fig. 7.7 The ratio of converging to a PQE of the proposed algorithm

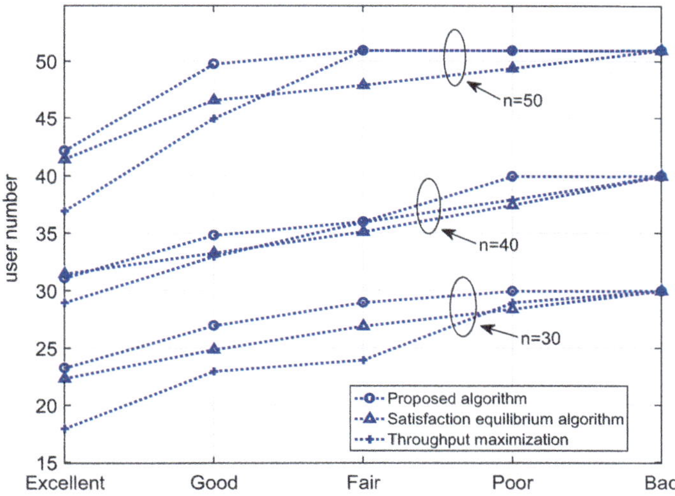

Fig. 7.8 Performance comparison of different algorithms with different user numbers

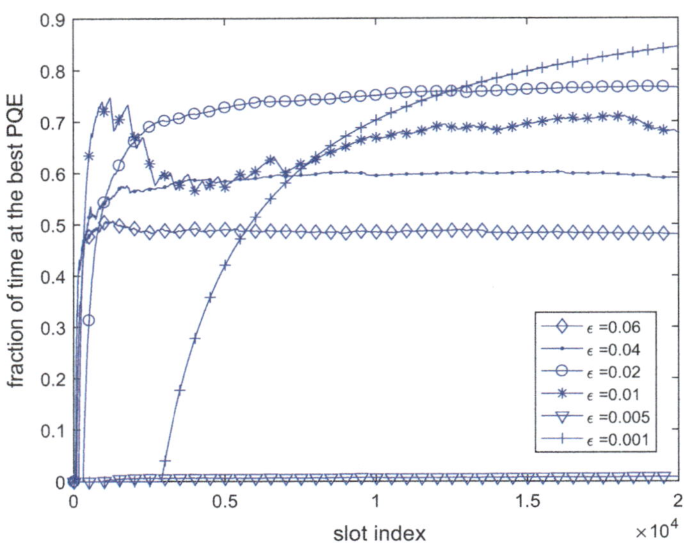

Fig. 7.9 Convergence performance with different experiment probability ε

7.6.2.3 TE-Based PQE Refinement Algorithm

For the TE-based PQE refinement algorithm, we set $c_5 = 0$ and $c_4 = 0.0001, , c_3 = 0.01$, $c_2 = 1$, $c_1 = 100$. Then the bound of the reassigned reward is

$$u_m \in [0, 100], \Delta u_m \in [-100, 100], \forall m \in \mathcal{M}.$$

Based on these parameters, we can obtain the state transition probability functions $G(\cdot)$ and $F(\cdot)$.

The experiment probability ε is a key factor affecting the system's stableness on equilibrium points. Figure 7.9 compares the convergence performance of best PQE of the proposed algorithm with different ε for $M' = 52$, $M = 16$ with equal user weight. The result indicates that a smaller ε can contribute to a larger fraction time at the equilibriums, but a relatively longer time to reach the stable state when $0.005 < \varepsilon < 0.1$. However, for the case of $\varepsilon = 0.005$, its performance becomes poor. Hence, it seems that there is no significant rule on the experiment probability changes, thus, the optimal experiment probability depends on experience.

Finally, we compare the algorithm with three algorithms. The first one is the so-called best response, where at each iteration, one randomly selected user selects one network to get a larger throughput. This algorithm is similar to the algorithm in [3]. The second algorithm is a Q learning algorithm, where all users adjust their access networks according to a predefined decision-making rule simultaneously at each iteration. The algorithm can be found in [22]. The third one is the satisfaction equilibrium

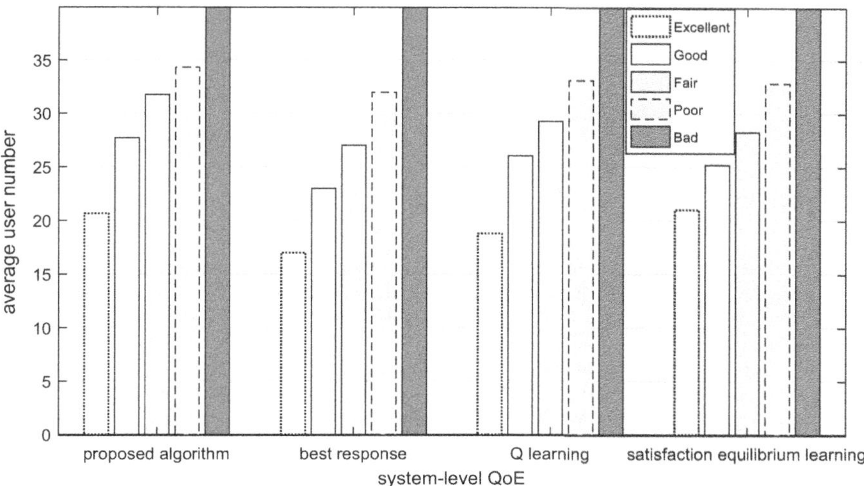

Fig. 7.10 Performance comparison with different algorithms

learning algorithm in [8], where each user stops changing the selected network once the "Excellent" level QoE is arrived, otherwise, it randomly accesses one network. Note that the former two algorithms represent existing throughput-centric methods, while the third algorithm only differentiates "satisfaction" and "unsatisfaction" states.

Figure 7.10 shows the performance comparison result for $M = 40$ with equal user weight. The proposed algorithm possesses performance gain to all the other algorithms except for satisfaction equilibrium learning algorithm. Specifically, the number of users with "Excellent" level is slightly smaller than that of the satisfaction equilibrium learning algorithm, while the numbers of users with no worse performance than "Good", "Fair", and "Poor" are larger. In addition, we compare the Jain's indexes and find out that the proposed algorithm achieves better fairness performance than the satisfaction equilibrium learning algorithm. The above results indicate that the TE-based PQE refinement algorithm achieves much better performance, which could help us to exploit the user demand diversity in distributed learning.

7.7 Conclusion

This chapter studies how to improve system efficiency in distributed network selection for multiple user cases with heterogeneous demand. The key insight is to exploit the user demand diversity. We propose a novel QoE game to analyze the system properties under the framework of user demand centric optimization. The properties of QoE game and QoE equilibrium are analyzed and the definition of user demand diversity gain is presented. A SLA-based MARL algorithm achieving PQE is proposed. To achieve user demand diversity gain, a TE-based MARL algorithm is proposed to converge to best PQEs.

References

1. Trestian R, Ormond O, Muntean G (2012) Game theory-based network selection: solutions and challenges. IEEE Commun Sruv Tut 14(4):1018–1044
2. Liu D et al (2016) User association in 5G networks: a survey and an outlook. IEEE Commun Sruv Tut 18(2):1018–1044
3. Keshavarz-Haddad A, Aryafar E, Wang M, Chiang M (2017) HetNets selection by clients: convergence, efficiency, and practicality. IEEE ACM Trans Netw 25(1):406–419
4. Zhu K, Niyato D and Ping W (2010) Network selection in heterogeneous wireless networks: evolution with incomplete information. In: IEEE wireless communications and networking conference (WCNC)
5. Zhu K, Hossain E, Niyato D (2014) Pricing, spectrum sharing, and service selection in two-tier small cell networks: a hierarchical dynamic game approach. IEEE Trans Mob Comput 13(8):1843–1856
6. Du Z, Wu Q et al (2015) Exploiting user demand diversity in heterogeneous wireless networks. IEEE Trans Wirel Commun 14(8):4142–4155
7. Deb S, Nagaraj K, Srinivasan V (2011) MOTA: engineering an operator agnostic mobile service. MobiCom (2011)
8. Perlaza SM, Tembine H, Lasaulce S et al (2011) Quality-of-service provisioning in decentralized networks: a satisfaction equilibrium approach. IEEE J-STSP 6(2):104–116
9. Rakocevic V, Griffiths J, Cope G (2001) Performance analysis of bandwidth allocation schemes in multiservice IP networks using utility functions. In: Proceedings of the 17th international teletraffic congress (ITC)
10. Reis AB, Chakareski J, Kassler A et al (2010) Distortion optimized multi-service scheduling for next-generation wireless mesh networks. In: IEEE INFOCOM
11. Kelly FP (1997) Charging and rate control for elastic traffic. Eur Trans Telecommun 8:33–37
12. Monderer D, Sharpley LS (1996) Potential games. Games Econ Behav 14:124–143
13. Milchtaich I (2009) Weighted congestion games with separable preferences. Game Econ Behav 67:750–757
14. Mavronicolas M, Milchtaich I et al (2007) Congestion games with player-specific constants. In: International symposium on mathematical foundations of computer science (MFCS)
15. Sastry P, Phansalkar V, Thathachar M (1994) Decentralized learning of nash equilibria in multi-person stochastic games with incomplete information. IEEE Trans Syst Man Cybern B 24(5):769–777
16. Xu Y, Wang J, Wu Q et al (2012) Opportunistic spectrum access in unknown dynamic environment: a game-theoretic stochastic learning solution. IEEE Trans Wireless Commun 11(4):1380–1391
17. Pradelski BS, Young HP (2010) Efficiency and equilibrium in trial and error learning. University of Oxford, Department of Economics, Economics Series Working Papers
18. Coucheney P, Touati C, Gaujal B (2009) Fair and efficient user-network association algorithm for multi-technology wireless networks. In: IEEE INFOCOM
19. Xue P, Gong P, Park J et al (2012) Radio resource management with proportional rate constraint in the heterogeneous networks. IEEE Trans Wirel Commun 11(3):1066–1075
20. 3GPP TR 36.814 V9.0.0 (2010-03)
21. How much bandwidth does Skype need? https://support.skype.com/en
22. Niyato D, Hossain E (2009) Dynamics of network selection in heterogeneous wireless networks: an evolutionary game approach. IEEE Trans Veh Technol 58(4):2008–2017 (2009)

Chapter 8
Future Work

Following the idea of this book, we would like to envision some research trends for user-centric online network selection optimization, which may not be limited to network selection but general resource management problems in wireless networks.

8.1 Personalized QoE Optimization

As the introduction of three typical service types eMBB (enhanced Mobile Broadband), URLLC (Ultra-Reliable Low-Latency Communications), and mMTC (massive Machine-Type Communications), today's mobile system is evolving toward more fine-grained service provisioning. It will be a natural way to provide customized QoE for users in the future. While QoE has received much attention in the community, there is a long road to realize personalized QoE optimization. Current QoE optimization is general in the sense that little personal information or user-specific profile is considered. In most cases, only traffic-dependent QoE utility is used as the optimization objective, with little consideration on the problem modeling process itself. The context-aware solution in Chap. 5 provides a good idea, i.e., taking context-specific information into network selection optimization, but the context resolution is not sufficient and more information remains to be mined and utilized.

8.2 New QoE Models

Following the above-mentioned personalized QoE optimization trend, new QoE models are needed. Currently available QoE models are too limited to satisfy future requirements. As new applications are increasingly emerging in mobile Internet, the

© Springer Nature Singapore Pte Ltd. 2020

Z. Du et al., *Towards User-Centric Intelligent Network Selection
in 5G Heterogeneous Wireless Networks*,
https://doi.org/10.1007/978-981-15-1120-2_8

development of function model faces challenges. Specifically, there is a lack of QoE function models for more interactive and immersive services such as augmented reality, virtual reality, and ultra-high-definition (UHD) 3D video. In addition, since smartphones nowadays can run multiple applications at the same time [1] (e.g., web browsing, video streaming, and file download), sophisticated QoE function models are needed to model the compound application. Our recent research in [2, 3] tries to model the dynamic QoE in terms of second-order QoE. On the other hand, recent advances on model-free QoE models and approaches have proved some interesting results. These approaches just rely on user behaviors or feedback to mine QoE and even guide resource optimization. For examples, the click of a dedicated button is used to capture users' perceptions whenever they feel dissatisfied with the quality of the application in [4]. The speaker' s attempt to recover lost speech using keywords or phrases, like "sorry?" and "hello?" in the VOIP call, could be used to infer QoE [5]. The recent work even tries to predicate QoE from facial expression and gaze direction [6].

8.3 Deep RL Based Solutions

To realize fine-grained resource management, we need to promote the model resolution, which inevitably leads to increasingly complex RL model. That is, the state space and action space in the underlying MDP will become large and even high-dimensional and the number of involving learning agents in game theory could become large. The increased model complexity will lower the algorithm efficiency and even makes algorithms intractable. Thanks to recent advances on deep learning, deep reinforcement learning (DRL) [7] has been proposed to tackle large-scale RL problem. DRL employs deep neural networks to predicate value functions for MDP based on past experience, successfully handling the learning efficiency issue for large-scale model. Accordingly, we have seen that DRL has been applied in network handover [8] and other resource management problems such as power control and spectrum sharing. While DRL has opened new doors for high-dimensional resource management, there are still some issues to be solved. The biggest issue is cost in training deep neural networks. It is reported that due to the large parameter space, the large number of operations and memory access tasks during both training and inference stages of deep neural networks will incur high intense computation burden and energy cost [9, 10]. Meanwhile, the training process could be impractical for resource management. Note that the training of deep neural network requires a large number of samples. Simulating these samples in computer game is easy, but it is impossible to accumulate the samples from practical systems running for resource management. Therefore, efficient DRL algorithms are needed.

References

1. Baker M (2012) From LTE-advanced to the future. IEEE Commun Mag 50(2):116–120
2. Du Z et al (2019) Second-order multi-armed bandit learning for online optimization in communication and networks. In: Proceedings of the ACM turing celebration conference-China (ACM TURC)
3. Du Z, et al (2019) Second-order reinforcement learning for end-to-end online path selection with QoE dynamics, submitted
4. Chen KT, Tu CC, Xiao WC (2009) Oneclick: a framework for measuring network quality of experience. In: IEEE INFOCOM
5. Hassan JA, Hassan M et al (2012) Managing quality of experience for wireless VOIP using noncooperative games. IEEE J Sel Areas Commun 30(7):1193–1204
6. Porcu S, Floris A, Atzori L (2019) Towards the prediction of the quality of experience from facial expression and gaze direction. ICIN
7. Mnih V, Kavukcuoglu K et al (2015) Human-level control through deep reinforcement learning. Nature 518:529–533
8. Wang Z, Li L et al (2018) Handover control in wireless systems via asynchronous multiuser deep reinforcement learning. IEEE Internet Things J 5(6):4296–4307
9. Schwartz R et al (2019) Green AI (2019). arXiv:1907.10597v2
10. Yang TJ et al (2017) Designing energy-efficient convolutional neural networks using energy-aware pruning. CVPR

Index

© Springer Nature Singapore Pte Ltd. 2020
Z. Du et al., *Towards User-Centric Intelligent Network Selection
in 5G Heterogeneous Wireless Networks*,
https://doi.org/10.1007/978-981-15-1120-2

The manufacturer's authorised representative in the EU is Springer
Nature Customer Service Centre GmbH, Europaplatz 3, 69115 Heidelberg,
Germany. If you have any concerns regarding our products, please
contact ProductSafety@springernature.com

Printed and bound by CPI Group (UK) Ltd, Croydon, CR0 4YY
28/11/2025
02007732-0003